CAMBRIDGE STUDIES IN ADVANCED MATHEMATICS 20

Introductory lectures on Siegel modular forms

Already published

1 W.M.L. Holcombe *Algebraic automata theory*
2 K. Petersen *Ergodic theory*
3 P.T. Johnstone *Stone spaces*
4 W.H. Schikhof *Ultrametric calculus*
5 J-P. Kahane *Some random series of functions*, second edition
6 H. Cohn *Introduction to the construction of class fields*
7 J. Lambek & P.J. Scott *Introduction to higher-order categorical logic*
8 H. Matsumura *Commutative ring theory*
9 C.B. Thomas *Characteristic classes and the cohomology of finite groups*
10 M. Aschbacher *Finite group theory*
11 J.L. Alperin *Local representation theory*
12 P. Koosis *The logarithmic integral: 1*
13 A. Pietsch *Eigenvalues and s-numbers*
14 S.J. Patterson *Introduction to the theory of the Riemann zeta-function*
15 H-J. Baues *Algebraic homotopy*
16 V.S. Varadarajan *Introduction to harmonic analysis on semisimple Lie groups*
17 W. Dicks & M.J. Dunwoody *Groups acting on graphs*
18 L.J. Corwin & F. P. Greenleaf *Representations of nilpotent Lie groups and their applications*
20 H. Klingen *Introduction to modular functions*
22 M. Collins *Representations and characters of finite groups*

Introductory lectures on Siegel modular forms

HELMUT KLINGEN

Professor of Mathematics
University of Freiburg

The right of the University of Cambridge to print and sell all manner of books was granted by Henry VIII in 1534. The University has printed and published continuously since 1584.

CAMBRIDGE UNIVERSITY PRESS

Cambridge

New York Port Chester Melbourne Sydney

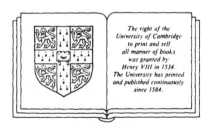

Published by the Press Syndicate of the University of Cambridge
The Pitt Building, Trumpington Street, Cambridge CB2 1RP
40 West 20th Street, New York, NY 10011, USA
10 Stamford Road, Oakleigh, Melbourne 3166, Australia

First published 1990

British Library cataloguing in publication data
Klingen, Helmut
Introductory lectures on Siegel modular forms.
1. Mathematics. Automorphic functions
I. Title
515.7

Library of Congress cataloguing in publication data
Klingen, Helmut.
Introductory lectures on Siegel modular forms/Helmut Klingen.
p. cm. – (Cambridge studies in advanced mathematics; 20)
Bibliography: p.
Includes index.
ISBN 0 521 35052 2
1. Siegel domains. 2. Modular groups. I. Title. II. Series.
QA331.K615 1990 515–dc20 89-9740 CIP

ISBN 0 521 35052 2

Transferred to digital printing 2003

AO

To my family,
Anita, Christoph and Philipp

Contents

Preface

The theory of automorphic functions in one complex variable was created during the second half of the nineteenth and the beginning of the twentieth centuries. Important contributions are due to such illustrious mathematicians as F. Klein, P. Koebe and H. Poincaré. Two sources may be traced: the uniformization theory of algebraic functions, and certain topics in number theory. Automorphic functions with respect to groups with compact quotient space on the one hand and elliptic modular functions on the other are examples of these two aspects. In several complex variables there is no analogue of uniformization theory; the class of automorphic functions which can be considered becomes much narrower, and the underlying groups are, in general, arithmetically defined.

In the mid-1930s C.L. Siegel discovered a new type of automorphic forms and functions in connection with his famous investigations on the analytic theory of quadratic forms. He denoted these functions as 'modular functions of degree *n*'; nowadays they are called 'Siegel modular functions'. Next to Abelian functions they are the most important example of automorphic functions in several complex variables, and they very soon became a touchstone to test the efficiency of general methods in several complex variables and other fields. Only recently, Hilbert modular functions have achieved a similar position due to the progress made in that area by K. Doi, F. Hirzebruch, F.W. Knöller, H. Naganuma and D. Zagier, amongst others. Siegel himself developed many powerful methods, and a steadily growing group of mathematicians increased the knowledge of Siegel modular forms, or found similar types of functions such as Hermitian modular forms, Hilbert–Siegel modular forms, or recently modular forms on half-spaces of quaternions. The need for a unified comprehensive treatment of these different but related theories became obvious. This was realized within the framework of arithmetically defined subgroups of algebraic groups and corresponding automorphic functions by the impressive work of W.L. Baily, A. Borel, R.P. Langlands and I.I. Pjateckij-Šapiro around 1965. Since then, further progress has been achieved only partly in this generality, but also in much more concrete situations, for instance A.N. Andrianov's results on Hecke's theory or the work of E. Freitag, D. Mumford and Y.-S. Tai on the structure of the function fields.

This book consists of lectures on Siegel modular forms that I have delivered in steadily improved versions, first in 1968 at the Tata Institute of Fundamental Research in Bombay, and afterwards on several occasions at the University of Freiburg. The audience in mind was composed of students who had taken only a one-complex-variable course besides having some basic knowledge in algebra, number theory and topology. In particular, in order to understand this book, the reader needs no knowledge of several complex variables, except perhaps the concept of a holomorphic or meromorphic function. The lectures were designed for a one-semester course with the intention of offering an easily accessible partial survey of the elementary parts of an exceptionally active field in mathematics. Consequently the selection of topics can by no means claim to be complete; even essential subjects like Satake's compactification or Hecke's theory cannot be included in this book. We restrict ourselves to the full modular group neglecting technical difficulties arising from subgroups. The reader, however, should feel encouraged to deal with the more advanced parts of the theory afterwards, using other books or the original literature recommended in the text.

Formulas, theorems etc. are numbered separately in each section. If a reference to a previous section is cited, the number of the section is placed in front of the number of the formula, theorem etc. For example, (4.1) means formula (1) of §4.

Acknowledgements are due to Siegfried Böcherer and Petra Ploch for their valuable comments and detailed reading of the manuscript. I am much indebted to Ruth Müller for her careful typing of the manuscript and her patience. Finally, I would like to take the opportunity to thank Cambridge University Press for the invitation and encouragement to write this book.

Freiburg, 1988 H. Klingen

I

The modular group

1 The symplectic group

The symplectic group over $\mathbb{R}$ is a subgroup of the general linear group, defined by certain algebraic equations, and appears from there as an algebraic group.

Definition 1

Let n be a positive integer. The symplectic group of degree n over $\mathbb{R}$ is the subgroup

$$Sp(n, \mathbb{R}) = \{m \in GL(2n, \mathbb{R}) \mid j[m] = j\}, \qquad j = \begin{pmatrix} 0 & \mathbf{1} \\ -\mathbf{1} & 0 \end{pmatrix}.$$

Here j is decomposed into $n \times n$ blocks, consisting of the identity matrix $\mathbf{1}$ and the zero matrix 0, respectively. The brackets [ ] always mean the transformation $a[b] := {}^t bab$. The set $Sp(n, \mathbb{R})$ is closed with respect to the group operations in $GL(2n, \mathbb{R})$, since clearly $\mathbf{1} \in Sp(n, \mathbb{R})$, $j[m] = j$ is equivalent to $j = j[m^{-1}]$ by transforming with m^{-1}, and finally $j[m_\nu] = j$, $\nu = 1, 2$, imply

$$j[m_1 m_2] = j[m_1][m_2] = j[m_2] = j.$$

Decomposing m into $n \times n$ blocks,

$$m = \begin{pmatrix} a & b \\ c & d \end{pmatrix},$$

the condition $j[m] = j$ is seen to be equivalent to

$${}^t ac,\ {}^t bd \quad \text{symmetric}, \qquad {}^t ad - {}^t cb = \mathbf{1}$$

or, evaluating $j[{}^t m] = j$, to

$$a\,{}^t b,\ c\,{}^t d \quad \text{symmetric}, \qquad a\,{}^t d - b\,{}^t c = \mathbf{1}.$$

Both conditions separately characterize the elements m of $Sp(n, \mathbb{R})$ completely. As a consequence we obtain the following formula for the inverse of a symplectic matrix:

$$m^{-1} = \begin{pmatrix} {}^t d & -{}^t b \\ -{}^t c & {}^t a \end{pmatrix}.$$

For $n = 1$, symplecticity just means $\det m = 1$. In general, $\det m = \pm 1$ can be derived from the definitions for arbitrary $m \in Sp(n, \mathbb{R})$. In fact $\det m = 1$ is true, but the converse obviously does not hold for $n > 1$.

It is well known how $Sp(1, \mathbb{R})$ acts on the upper half-plane as a group of biholomorphic mappings. To generalize this subject to arbitrary n, Siegel's half-space of degree n will be introduced.

Definition 2

Let n be a positive integer. Siegel's half-space of degree n consists of all n-rowed complex symmetric matrices z, the imaginary part of which is positive definite,

$$H_n = \{z = x + iy \mid {}^t z = z, y > 0\}.$$

Considering the independent entries z_{kl} $(k \leq l)$ of z as coordinates, H_n becomes an open subset of $\mathbb{C}^{n(n+1)/2}$. Since for positive matrices y_1, y_2 and real λ,

$$\lambda y_1 + (1 - \lambda) y_2 > 0 \qquad (0 \leq \lambda \leq 1),$$

H_n turns out to be a convex domain in $\mathbb{C}^{n(n+1)/2}$. As a convex domain, H_n in particular is simply connected. Now we have to use the concept of a holomorphic function in several complex variables for the first time. Following K. Weierstraß we may introduce holomorphic functions as complex-valued functions which can be represented locally by power series expansions. Holomorphic mappings between two domains embedded in complex number spaces are described by holomorphic coordinate-functions. Then we may state

Proposition 1

$Sp(n, \mathbb{R})$ acts on H_n as a group of biholomorphic automorphisms by

$$Sp(n, \mathbb{R}) \times H_n \to H_n, \qquad (m, z) \mapsto m\langle z \rangle := (az + b)(cz + d)^{-1}.$$

Proof

First we show that the matrix $cz + d$ is non-singular. Let $z \in H_n$, $m \in Sp(n, \mathbb{R})$ and put

$$p = az + b, \qquad q = cz + d.$$

Using the symplecticity of m we obtain

$$\begin{aligned} {}^t p \bar{q} - {}^t q \bar{p} &= (z\,{}^t a + {}^t b)(c\bar{z} + d) - (z\,{}^t c + {}^t d)(a\bar{z} + b) \\ &= z - \bar{z} = 2iy. \end{aligned} \tag{1}$$

Now let ξ be any n-rowed complex column such that $q\xi = 0$. Then ${}^t\xi y\bar{\xi} = 0$ by the equation above, hence $\xi = 0$ since y is positive. The system of linear equations $q\xi = 0$ has only the trivial solution; therefore q is non-singular and $m\langle z\rangle$ is well defined. Next, the symmetry of $z^* = m\langle z\rangle$ is equivalent to

$${}^tpq = ({}^tz\,{}^ta + {}^tb)(cz + d) = ({}^tz\,{}^tc + {}^td)(az + b) = {}^tqp,$$

which follows again from the symplecticity of m and the symmetry of z. Now we infer from (1) for the imaginary part of z^*

$$y^* = \frac{1}{2i}(z^* - \bar{z}^*) = \frac{1}{2i}({}^tp\bar{q} - {}^tq\bar{p})\{\bar{q}^{-1}\} = y\{\bar{q}^{-1}\} > 0.$$

Here and from now on we use the brackets $\{\ \}$ to denote the transformation as a Hermitian form, i.e. $a\{b\} := {}^t\bar{b}ab$. So $z^* \in H_n$ and the map in question is of the indicated kind. It is an action of $Sp(n, \mathbb{R})$ on H_n, since one verifies immediately that

$$(m_1 m_2)\langle z\rangle = m_1\langle m_2\langle z\rangle\rangle, \qquad e\langle z\rangle = z$$

for $m_1, m_2 \in Sp(n, \mathbb{R})$, $z \in H_n$ and the unit element e of $Sp(n, \mathbb{R})$. As with every action of a group on a set, the corresponding mappings

$$H_n \to H_n, \qquad z \mapsto m\langle z\rangle \tag{2}$$

for fixed $m \in Sp(n, \mathbb{R})$ are bijective. These 'symplectic maps' are holomorphic, since they are rational, and biholomorphic, as the inverse map is performed with m^{-1}. Of course, we could as well have used the fact that every holomorphic bijective map of a domain onto itself is biholomorphic.

If we assign to each $m \in Sp(n, \mathbb{R})$ the automorphism (2) of H_n, we obtain a group homomorphism

$$Sp(n, \mathbb{R}) \to \mathrm{Bihol}(H_n)$$

of the symplectic group into the group of biholomorphic automorphisms of H_n. The kernel of this homomorphism consists of $\pm\mathbf{1}$, since the identity

$$m\langle z\rangle = z$$

in H_n implies

$$az + b = z(cz + d)$$

for arbitrary symmetric z, hence $a = d = \pm\mathbf{1}$, $b = c = 0$. Even the surjectivity of the homomorphism can be shown [64] using a generalization of Schwarz's lemma. So the analogy with the one-variable case is remarkable.

Nevertheless, for certain purposes it is useful to have available another realization of H_n, namely as a bounded symmetric domain in the sense of E. Cartan. This concept was created in order to guarantee the existence of enough non-trivial biholomorphic automorphisms of a domain. Note that in several complex variables there exist domains for which the identity is the only biholomorphic automorphism. Such a domain would be of no use in the theory of automorphic functions. A domain is called homogeneous if the group of biholomorphic automorphisms acts transitively; it is called symmetric if to each point there exists an involution in the group of biholomorphic automorphisms with the given point as a single fixed point. E. Cartan [13] proved that each bounded symmetric domain is homogeneous. I.I. Pjateckij-Šapiro [55] discovered the first example of a non-symmetric but homogeneous bounded domain in 1959. Bounded symmetric domains were classified by E. Cartan [13]; the larger class of bounded homogeneous domains was investigated by I.M. Gelfand, S.G. Gindekin, I.I. Pjateckij-Šapiro and E.B. Vinberg. E. Cartan obtained four main types of irreducible bounded symmetric domains and two exceptional ones that appear only for dimensions 16 and 27, respectively. Here irreducible means that the domain cannot be decomposed into the product of two domains of the same kind. One of Cartan's main types is relevant for our considerations, it is a generalization of the unit-circle to several complex variables.

Definition 3

Let n be a positive integer. The unit-circle of degree n consists of all n-rowed complex symmetric matrices w, for which the Hermitian matrix $\mathbf{1}-w\bar{w}$ *is positive definite,*

$$D_n = \{w \mid {}^t w = w, \mathbf{1} - w\bar{w} > 0\}.$$

Clearly D_n is a bounded domain in $\mathbb{C}^{n(n+1)/2}$. It is related to H_n by a generalized Cayley transformation.

Proposition 2

The Cayley transformation

$$l: H_n \to D_n, \qquad z \mapsto w := l\langle z\rangle = (z - i\mathbf{1})(z + i\mathbf{1})^{-1}$$

maps H_n *biholomorphically onto* D_n.

Proof

For $z \in H_n$ we know that $\det z \neq 0$ (a special case of $\det(cz + d) \neq 0$ in Proposition 1). Since $z + i\mathbf{1} \in H_n$, the Cayley transformation is well defined. Now the symmetry of z implies the symmetry of w and

$$\begin{aligned}\mathbf{1} - w\bar{w} &= \mathbf{1} - (z + \mathrm{i}\mathbf{1})^{-1}(z - \mathrm{i}\mathbf{1})(\bar{z} + \mathrm{i}\mathbf{1})(\bar{z} - \mathrm{i}\mathbf{1})^{-1} \\ &= ((z + \mathrm{i}\mathbf{1})(\bar{z} - \mathrm{i}\mathbf{1}) - (z - \mathrm{i}\mathbf{1})(\bar{z} + \mathrm{i}\mathbf{1}))\{(\bar{z} - \mathrm{i}\mathbf{1})^{-1}\} \\ &= 4y\{(\bar{z} - \mathrm{i}\mathbf{1})^{-1}\} > 0.\end{aligned}$$

Thus the Cayley transformation maps H_n into D_n. On the other hand, $\mathbf{1} - w$ is non-singular for arbitrary $w \in D_n$, for $(\mathbf{1} - w)\xi = 0$ implies $(\mathbf{1} - w\bar{w})\{\xi\} = 0$, hence $\xi = 0$. We may therefore form

$$z = \mathrm{i}(\mathbf{1} + w)(\mathbf{1} - w)^{-1} \tag{3}$$

for $w \in D_n$. The symmetry of w implies the symmetry of z; furthermore

$$\begin{aligned}y &= \tfrac{1}{2}((\mathbf{1} - w)^{-1}(\mathbf{1} + w) + (\mathbf{1} + \bar{w})(\mathbf{1} - \bar{w})^{-1}) \\ &= \tfrac{1}{2}((\mathbf{1} + w)(\mathbf{1} - \bar{w}) + (\mathbf{1} - w)(\mathbf{1} + \bar{w}))\{(\mathbf{1} - \bar{w})^{-1}\} \\ &= (\mathbf{1} - w\bar{w})\{(\mathbf{1} - \bar{w})^{-1}\} > 0.\end{aligned}$$

Therefore D_n is mapped into H_n by (3). Obviously the Cayley transformation and (3) are inverse to each other, hence the proposition is proved.

If we introduce the $2n \times 2n$ matrix

$$l = \begin{pmatrix} \mathbf{1} & -\mathrm{i}\mathbf{1} \\ \mathbf{1} & \mathrm{i}\mathbf{1} \end{pmatrix}$$

associated with the Cayley transformation, then the subgroup

$$\Phi_n = l\, Sp(n, \mathbb{R}) l^{-1}$$

of $GL(2n, \mathbb{C})$ acts in the same manner on the unit-circle D_n as $Sp(n, \mathbb{R})$ did on the half-space H_n. Let us determine explicitly the conditions that characterize Φ_n as a subgroup of $GL(2n, \mathbb{C})$. For $m \in GL(2n, \mathbb{C})$ we may state the equivalence

$$m \in Sp(n, \mathbb{R}) \quad \Leftrightarrow \quad j\{m\} = j, \qquad m = \bar{m}.$$

Now we put $m^* = lml^{-1}$ and consider each condition on the right separately:

$$j\{m\} = j \quad \Leftrightarrow \quad j\{l^{-1}\}\{m^*\} = j\{l^{-1}\} \quad \Leftrightarrow \quad k\{m^*\} = k$$

with

$$k = \begin{pmatrix} \mathbf{1} & 0 \\ 0 & -\mathbf{1} \end{pmatrix}$$

and

$$m = \bar{m} \quad \Leftrightarrow \quad m^* l\bar{l}^{-1} = l\bar{l}^{-1}\bar{m}^* \quad \Leftrightarrow \quad c^* = \bar{b}^*, d^* = \bar{a}^*$$

with

$$m^* = \begin{pmatrix} a^* & b^* \\ c^* & d^* \end{pmatrix}.$$

Therefore we obtain

$$\Phi_n = \left\{ m \in GL(2n, \mathbb{C}) \mid m = \begin{pmatrix} a & b \\ \bar{b} & \bar{a} \end{pmatrix}, k\{m\} = k \right\}. \tag{4}$$

As an illustration we verify directly

Proposition 3

$Sp(n, \mathbb{R})$, respectively Φ_n, act transitively on H_n, respectively D_n. Both domains are symmetric in the sense of E. Cartan.

Proof

Because of Proposition 2 it is sufficient to consider the action of Φ_n on D_n. First we show its transitivity. Let w be an arbitrary point of D_n; we have to look for an element $m \in \Phi_n$ which transforms w into any distinguished point, for instance the origin. Since $\mathbf{1} - w\bar{w}$ is positive, there exists an $n \times n$ matrix a with complex entries such that

$$(\mathbf{1} - w\bar{w})\{a\} = \mathbf{1}.$$

Put

$$m = \begin{pmatrix} a & b \\ \bar{b} & \bar{a} \end{pmatrix}, \qquad b = w\bar{a}.$$

Then we obtain

$${}^t\bar{a}a - {}^t b\bar{b} = \mathbf{1}, \qquad {}^t\bar{a}b \text{ symmetric},$$

which is equivalent to $k\{m\} = k$. Therefore $m \in \Phi_n$ by (4) and obviously $w = m\langle 0 \rangle$. Concerning the symmetry, note that $w \mapsto -w$ is an involution of D_n which corresponds to an element of Φ_n and has the origin as a single fixed point. Since we already know that Φ_n is transitive we obtain the same property for any other point of D_n.

On this occasion we prove a useful lemma, which allows us to strengthen Proposition 3 as a corollary.

Lemma

Let w be any n-rowed complex symmetric matrix and

$$d = \begin{pmatrix} \lambda_1^{1/2} & & 0 \\ & \ddots & \\ 0 & & \lambda_n^{1/2} \end{pmatrix},$$

where $\lambda_1, \ldots, \lambda_n$ are the eigenvalues of the Hermitian matrix $w\bar{w}$ in any prescribed order. Then there exists a unitary matrix u such that $w = d[u]$.

Proof

The argument becomes a little involved since transformations as quadratic forms (denoted by [ ]) and as Hermitian forms (denoted by { }) appear at the same time. If w is a diagonal matrix with elements $w_1, \ldots, w_n$ in the main diagonal, we may assume

$$\lambda_\nu = w_\nu \bar{w}_\nu \qquad (\nu = 1, \ldots, n)$$

after reordering, which is an orthogonal transformation. Then we take for u the diagonal matrix formed with the elements

$$\left(\frac{w_\nu}{\bar{w}_\nu}\right)^{1/4} \qquad (\nu = 1, \ldots, n)$$

or 1, if w_ν vanishes. For arbitrary w first determine a unitary matrix u_1 such that $w\bar{w}\{u_1\} = d^2$. Then $q = w[\bar{u}_1]$ is symmetric and $q\bar{q} = d^2$ real. Therefore the real and the imaginary part of q commute and can be transformed into a diagonal form simultaneously by an orthogonal matrix u_2. Then $q[u_2] = w[\bar{u}_1 u_2]$ becomes diagonal too and $\bar{u}_1 u_2$ is unitary. So we have transformed w as a quadratic form into a diagonal matrix by the unitary matrix $u = \bar{u}_1 u_2$. Since such a transformation does not affect the eigenvalues of $w\bar{w}$, we are back to the first case.

Corollary

Let w_1, w_2 be two arbitrary points in D_n. Then there exists an $m \in \Phi_n$ which simultaneously transforms w_1 into 0 and w_2 into a diagonal matrix t, the diagonal elements of which satisfy $0 \leq t_1 \leq \cdots \leq t_n < 1$.

By Proposition 3 we may assume $w_1 = 0$. Take for $t_1, \ldots, t_n$ the square roots of the eigenvalues of $w_2 \bar{w}_2$. Then the lemma yields a unitary matrix u such that $w_2 = t[u]$; the mapping $w \mapsto w[u^{-1}]$ is induced by the action of Φ_n and has the desired properties.

The idea of considering the generalized unit-circle instead of Siegel's half-space has two advantages. First, the boundedness is often good for topological considerations, and second the action of Φ_n may be extended holomorphically onto the closure $\bar{D}_n$ of the unit-circle.

Proposition 4

The mappings

$$w \mapsto m\langle w\rangle = (aw + b)(\bar{b}w + \bar{a})^{-1} \qquad (m \in \Phi_n)$$

are topological automorphisms of $\bar{D}_n$. They are holomorphic on a neighborhood of $\bar{D}_n$ (depending on m).

Proof
It is sufficient to verify

$$\det(\bar{b}w + \bar{a}) \neq 0$$

for all $m \in \Phi_n$, $w \in \bar{D}_n$. Now let ξ be a complex column satisfying

$${}^t\xi(\bar{b}w + \bar{a}) = 0.$$

Then

$$a\,{}^t\bar{a}\{\xi\} = \bar{w}w\{{}^t\bar{b}\xi\}.$$

Note that the condition $k\{m\} = k$ in (4) is equivalent to $k\{{}^t\bar{m}\} = k$, i.e.

$$a\,{}^t\bar{a} - b\,{}^t\bar{b} = \mathbf{1}, \qquad a\,{}^tb \text{ symmetric.}$$

Thus we obtain

$$(\mathbf{1} + b\,{}^t\bar{b})\{\xi\} = \bar{w}w\{{}^t\bar{b}\xi\}$$

or

$$\mathbf{1}\{\xi\} + (\mathbf{1} - \bar{w}w)\{{}^t\bar{b}\xi\} = 0.$$

Since $\mathbf{1} - \bar{w}w \geq 0$, we infer $\xi = 0$.

Remark
For $Sp(n, \mathbb{R})$ the corresponding proposition is wrong as can be seen, for instance, from the map $z \mapsto -z^{-1}$.

In the remaining part of this section we will mention some basic facts of symplectic geometry which are used later. The well-known model of hyperbolic geometry in the upper half-plane can be generalized to Siegel's half-space H_n by introducing a certain invariant Riemannian metric. The geometrical properties were investigated originally by L.K. Hua, C.L. Siegel and M. Sugawara, and more recently by S. Helgason, A. Korányi, O. Loos, S. Murakami and J. Wolf, amongst others. Differentiating the symplectic mapping

$$z^* = (az + b)(cz + d)^{-1},$$

we obtain

$$\mathrm{d}z^*(cz + d) + z^*c\,\mathrm{d}z = a\,\mathrm{d}z$$

$$\mathrm{d}z^*[cz + d] + (z\,{}^ta + {}^tb)c\,\mathrm{d}z = (z\,{}^tc + {}^td)a\,\mathrm{d}z$$

$$\mathrm{d}z^*[cz + d] = \mathrm{d}z, \tag{5}$$

where $\mathrm{d}z = (\mathrm{d}z_{kl})$ denotes the matrix of the differentials $\mathrm{d}z_{kl}$. We easily check the transformation law for the imaginary part to be

$$y^* = y\{(cz + d)^{-1}\}. \tag{6}$$

From these two formulas we deduce immediately

$$\mathrm{d}z^*y^{*-1}\,\mathrm{d}\bar{z}^*y^{*-1} = {}^t(cz + d)^{-1}\,\mathrm{d}zy^{-1}\,\mathrm{d}\bar{z}y^{-1}\,{}^t(cz + d).$$

Hence the trace $\sigma(dzy^{-1}\,d\bar{z}y^{-1})$ defines a quadratic differential form in dx_{kl}, dy_{kl} $(k \le l)$ over $\mathbb{R}$, which is invariant with respect to the action of $Sp(n, \mathbb{R})$. Since this action is transitive by Proposition 3, we may check the positivity of the differential form at a single point, for instance $z = i\mathbf{1}$. But there

$$\sigma(dz\,d\bar{z}) = \sum_k (dx_{kk}^2 + dy_{kk}^2) + 2\sum_{k<l}(dx_{kl}^2 + dy_{kl}^2), \tag{7}$$

the positivity of which is obvious. Now we can introduce an invariant Riemannian metric on H_n by the symplectic line element

$$ds^2 = \sigma(dzy^{-1}\,d\bar{z}y^{-1}).$$

Let us transform this line element to the generalized unit-circle by Cayley's transformation. From

$$w = (z - i\mathbf{1})(z + i\mathbf{1})^{-1}$$

we deduce

$$dz = \frac{1}{2i}\,dw[z + i\mathbf{1}].$$

We have already used in the proof of Proposition 2 that

$$y = \tfrac{1}{4}(\mathbf{1} - w\bar{w})\{\bar{z} - i\mathbf{1}\}.$$

These two formulas allow the calculation of ds^2 in terms of w. The result is

$$ds^2 = 4\sigma(dw(\mathbf{1} - \bar{w}w)^{-1}\,d\bar{w}(\mathbf{1} - w\bar{w})^{-1}).$$

It is certainly interesting to study the Riemannian space H_n equipped with the symplectic metric ds^2 from a geometrical point of view. An excellent survey of the main results concerning geodesics, curvature etc. can be found in Siegel's paper [64]. Since we do not use these geometrical aspects in the following, we restrict ourselves to the determination of the symplectic volume element, which is important in integration theory.

In Riemannian geometry the invariant volume element is defined as the Euclidean volume element multiplied by the square root of the determinant of the quadratic differential form ds^2. Since $Sp(n, \mathbb{R})$ acts transitively on H_n, this volume element is uniquely determined by its invariance up to a constant factor. Therefore it is not necessary to compute the determinant of ds^2 explicitly if any invariant volume element is available by another argument. Then only the determination of an inessential factor remains open. To simplify the computation of Jacobians, note once and for all that the linear map

$$w \mapsto w[c]$$

from the space of n-rowed symmetric matrices w into itself has determinant $\det c^{n+1}$. From this observation all the Jacobians which appear in this book can be read off immediately. So we obtain from (5)

$$\det\left(\frac{\partial z^*}{\partial z}\right) = \det(cz + d)^{-n-1}$$

for the Jacobian of any symplectic map; or, introducing real coordinates,

$$\det\left(\frac{\partial(x^*, y^*)}{\partial(x, y)}\right) = \det\left(\frac{\partial(z^*, \bar{z}^*)}{\partial(z, \bar{z})}\right) = |\det(cz + d)|^{-2n-2}.$$

By (6) we have

$$\left(\frac{\det y^*}{\det y}\right)^{n+1} = |\det(cz + d)|^{-2n-2}.$$

So, if $\mathrm{d}x\,\mathrm{d}y = \prod_{k \le l} \mathrm{d}x_{kl}\,\mathrm{d}y_{kl}$ denotes the Euclidean volume element, then

$$\mathrm{d}v_n = \frac{\mathrm{d}x\,\mathrm{d}y}{\det y^{n+1}}$$

is invariant with respect to the action of $Sp(n, \mathbb{R})$. The volume element of the Riemannian metric $\mathrm{d}s^2$ differs from $\mathrm{d}v_n$ only by a constant factor, which is $2^{n(n-1)/2}$, as can be seen from (7). We call $\mathrm{d}v_n$ the symplectic volume element. A straightforward computation yields

$$\mathrm{d}v_n = 2^{n(n+1)} \frac{\mathrm{d}u\,\mathrm{d}v}{\det(1 - w\bar{w})^{n+1}} \qquad (w = u + iv)$$

for the volume element, after being carried over to the generalized unit-circle D_n by Cayley's transformation.

2 Minkowski's reduction theory

The imaginary parts of the points $z \in H_n$ form the subspace

$$P_n = \{y \mid y > 0\}$$

of $\mathbb{R}^{n(n+1)/2}$ consisting of all n-rowed positive definite matrices with real entries. It is an open convex subset of $\mathbb{R}^{n(n+1)/2}$. Moreover the ray originating from the point 0 and passing through any point $y \in P_n$ lies completely in P_n. Therefore P_n is a convex cone with vertex at the origin. Consider on the other hand the subgroup

$$\left\{m \in Sp(n, \mathbb{R}) \mid m = \begin{pmatrix} a & 0 \\ 0 & {}^t a^{-1} \end{pmatrix}, a \in GL(n, \mathbb{R})\right\}$$

of $Sp(n, \mathbb{R})$, which is canonically isomorphic to $GL(n, \mathbb{R})$. Then the action of $Sp(n, \mathbb{R})$ on H_n in Proposition 1.1 induces the action

$$GL(n, \mathbb{R}) \times P_n \to P_n, \qquad (a, y) \mapsto y[{}^t a]$$

of $GL(n,\mathbb{R})$ on P_n. Because of this connection we prefer to let $GL(n,\mathbb{R})$ operate on P_n from the left. If one assigns to $a \in GL(n,\mathbb{R})$ the map

$$P_n \to P_n, \qquad y \mapsto y[{}^t a],$$

one obtains a group homomorphism of $GL(n,\mathbb{R})$ into the group $\mathrm{Bij}(P_n)$ of bijective maps of P_n. The kernel is determined easily. Inserting $y = \mathbf{1}$ into the identity

$$y = y[a] \qquad (y \in P_n)$$

we first see that a is orthogonal. Then from

$$ya = ay$$

for all symmetric y we infer a to be a multiple of the identity matrix. Therefore the kernel consists of $\pm\mathbf{1}$.

By the considerations above we have found a close connection between the action of $Sp(n,\mathbb{R})$ on H_n and the action of $GL(n,\mathbb{R})$ on P_n. The same situation holds for arithmetically defined subgroups of $Sp(n,\mathbb{R})$ and $GL(n,\mathbb{R})$, respectively. For instance, we could have taken $Sp(n,\mathbb{Z})$ and $GL(n,\mathbb{Z})$. Reduction theory is concerned with the action of $GL(n,\mathbb{Z})$ on P_n and therefore is expected to be a powerful instrument in the handling of the group $Sp(n,\mathbb{Z})$, which is Siegel's modular group. However, having thus found a good motivation for the subsequent investigations, one should point out that reduction theory was not invented for this auxiliary purpose. There are more important applications to the theory of quadratic forms and to geometry of numbers.

Henceforth we call $GL(n,\mathbb{Z})$ the unimodular group of degree n and denote it by U_n. It consists of all n-rowed matrices u with entries in $\mathbb{Z}$ and $\det u = \pm 1$, called unimodular matrices. The group operation is of course matrix multiplication. Our main objective will be the determination of fundamental sets for the group action

$$U_n \times P_n \to P_n, \qquad (u, y) \mapsto y[{}^t u] \tag{1}$$

of a particularly simple form. The main results on reduction theory are due to J.L. Lagrange for $n = 2$, L.A. Seeber for $n = 3$, Ch. Hermite and H. Minkowski [52] for general n. Other approaches to reduction theory were discovered by H. Weyl and G.F. Voronoï.

Roughly speaking, a fundamental set is an irreducible complete set of representatives for the orbits of the underlying group action. In our case the group action is described by (1), and two points $y, y^* \in P_n$ belong to the same orbit if and only if there exists a $u \in U_n$ such that $y^* = y[u]$. The orbits are nothing other than the equivalence classes of this equivalence relation on P_n. We try to determine a representative of each orbit by certain minimization conditions. For any point y in P_n first determine the integral column $u_1 \neq 0$ such that $y[u_1]$ becomes minimal. Since y is positive, this minimum is attained for finitely many u_1 which are primitive; fix one of those. After having already determined $u_1, \ldots, u_{k-1}$, let u_k run over all integral columns such that the $n \times k$ matrix

$$(u_1, \ldots, u_{k-1}, u_k)$$

is primitive and ${}^t u_{k-1} y u_k \geq 0$. Then minimize $y[u_k]$ under these conditions. Note that the second condition can always be satisfied if one replaces u_k by $-u_k$ if necessary. Furthermore we remark that a matrix with integral entries is called primitive iff it is complementary to a unimodular matrix, or equivalently iff the minors of maximal size are coprime. After n steps we obtain a unimodular matrix $u = (u_1, \ldots, u_n)$. Let us consider the conditions satisfied by

$$y^* = y[u].$$

With the kth step the admissible columns u_k were exactly the kth columns of

$$u\begin{pmatrix} \mathbf{1} & a \\ 0 & b \end{pmatrix},$$

where $\mathbf{1}$ stands for the identity matrix of size $k - 1$, a is an integral and b an $(n - k + 1)$-rowed unimodular matrix. Therefore we obtain

$$y^*[g] \geq y^*_{kk}$$

for all integral columns g, the last $n - k + 1$ elements $g_k, \ldots, g_n$ of which are coprime. Furthermore the second condition means $y^*_{k-1,k} \geq 0$.

Definition 1

Minkowski's reduced domain is the set

$$R_n = \{y \in P_n \mid y \textit{ satisfies } \text{(i) and (ii)}\},$$

where

(i) $y[g] \geq y_{kk}$ *for all integral g with* $(g_k, \ldots, g_n) = 1 \quad (1 \leq k \leq n)$,

(ii) $y_{k,k+1} \geq 0 \quad (1 \leq k \leq n - 1)$.

The conditions (i) and (ii) are called reduction conditions and the elements of R_n reduced in the sense of Minkowski.

In the following we call integral columns g with $(g_k, \ldots, g_n) = 1$ 'k-admissible' for short. Note that so far R_n has been described as a subset of P_n by infinitely many linear inequalities. Another useful remark is concerned with the following fact. If $y \in R_n$ and y^* is obtained from y by deleting the last $n - k$ rows and columns, then $y^* \in R_k$ $(k = 1, \ldots, n)$. This can immediately be seen from the reduction conditions.

We have proved that P_n is covered by the images of R_n under the action of U_n, i.e.

$$P_n = \bigcup_{u \in U_n} u(R_n).$$

It is not difficult to deduce from the reduction conditions that each point of P_n is covered by at most finitely many images of R_n; more precisely, the number

$$\#\{u \in U_n \mid y[u] \in R_n\} \tag{2}$$

is shown to be finite for any fixed $y \in P_n$. Indeed, we infer from $y[u] \in R_n$ by (i) that

$$y_{11} = y[uu^{-1}e_1] \geq y[u_1],$$

where e_ν ($\nu = 1, \ldots, n$) denotes the νth unit vector and u_1 is the first column of u. Hence there are only finitely many possible values for u_1. Now assume $y[u], y[v] \in R_n$ and let u and v coincide with their first $k-1$ columns. Then after replacing $y[u]$ by y we have

$$y[v] \in R_n, \qquad v = \begin{pmatrix} \mathbf{1} & * \\ 0 & r \end{pmatrix}$$

with the $(k-1)$-rowed identity matrix $\mathbf{1}$. The kth column of v^{-1} is k-admissible and we obtain from (i)

$$y_{kk} = y[vv^{-1}e_k] \geq y[v_k].$$

Therefore we have only finitely many choices for the kth column v_k of v. After n steps we obtain (2). But we cannot see from our argument whether the bound in (2) may be chosen independently of y. This is an important question, which we will follow up later.

Next we deduce some consequences of Minkowski's reduction conditions.

Proposition 1

For any $y \in R_n$ the following inequalities hold:

(i) $y_k \leq y_{k+1}$ $\quad (k = 1, \ldots, n-1)$,

(ii) $|2y_{kl}| \leq y_k$ $\quad (k < l)$,

(iii) *there exists a positive real number $c_1 = c_1(n)$ depending only on n such that*

$$\det y \leq \prod_{\nu=1}^{n} y_\nu \leq c_1 \det y.$$

First we should mention that we have suppressed the double index of the diagonal elements of y, writing y_k instead of y_{kk} for simplicity. The essential part of the proposition is the right-hand inequality in (iii), which is called Minkowski's inequality. We may describe the significance of the proposition by the following remark. The finitely many inequalities (i)–(iii) of the proposition are implications of the infinitely many reduction conditions in Definition 1. On the other hand, we will see later that we do not lose the essential properties of R_n if we replace Minkowski's reduction conditions by the inequalities of the proposition. Against this background the transition to finiteness is the important point. But note that linearity gets lost with Minkowski's inequality.

Concerning the proof, (i), (ii) and the left-hand inequality in (iii) are trivial. Indeed, (i) follows from the reduction conditions of the first kind for index k and the k-admissible column $g = e_{k+1}$; then we take the l-admissible columns $g = e_k \pm e_l$ in Minkowski's reduction conditions of index l to obtain

$$y[g] = y_k + y_l \pm 2y_{kl} \geq y_l,$$

which is exactly (ii). The left-hand inequality in (iii) has a simple and well-known geometrical interpretation. Write the positive matrix y as

$$y = {}^t a a$$

with a non-singular real matrix a and consider the parallelepiped in $\mathbb{R}^n$ spanned by the columns of a. Its volume is $|\det a|$. Then

$$\det y \leq \prod_{\nu=1}^{n} y_\nu$$

says that, among all parallelepipeds spanned by vectors of fixed lengths, the rectangular one has maximal volume.

To prepare for the proof of Minkowski's inequality we first consider Hermite's lemma on the minima of positive definite quadratic forms. The minima in question are

$$\mu(y) = \min_g y[g],$$

where g runs over all integral columns different from zero; y is n-rowed and positive definite. For positive real numbers t we have $\mu(ty) = t\mu(y)$ and $\det(ty) = t^n \det y$. Therefore it seems reasonable to compare $\mu(y)$ with $\det y^{1/n}$.

Lemma 1 (Ch. Hermite)

There exists a constant $c_2 = c_2(n)$ depending only on n such that

$$\mu(y) \leq c_2 \det y^{1/n}$$

for all real n-rowed positive definite matrices y. A possible choice is

$$c_2 = (\tfrac{4}{3})^{(n-1)/2}.$$

Proof

We argue by induction on n. Obviously the inequality holds for $n = 1$. Assume it is true for forms of degree $n - 1$ and let y be any n-rowed positive definite matrix. If the minimum $\mu(y)$ is attained for g, then g is primitive and can be completed to a unimodular matrix u. Replacing y by $y[u]$ affects neither $\mu(y)$ nor $\det y$. Therefore we may assume that $y[g]$ becomes minimal for $g = e_1$, so $\mu(y) = y_1$. Now use the elementary identity

$$y = \begin{pmatrix} p & q \\ {}^tq & s \end{pmatrix} = \begin{pmatrix} p & 0 \\ 0 & t \end{pmatrix}\left[\begin{pmatrix} 1 & p^{-1}q \\ 0 & 1 \end{pmatrix}\right], \qquad t = s - p^{-1}[q], \tag{3}$$

which is valid for arbitrary matrices decomposed into blocks of appropriate size, p being non-singular. Of course we observe behind this formula the well-known method of 'completion of squares'. Apply (3) to y decomposed of type $(1, n - 1)$. Then

$$y[g] = y_1\left(g_1 + \frac{y_{12}}{y_1}g_2 + \cdots + \frac{y_{1n}}{y_1}g_n\right)^2 + \sum_{k,l>1} t_{kl}g_kg_l. \tag{4}$$

By the induction hypothesis there exists a non-trivial integral solution $g_2, \ldots, g_n$ of

$$\sum_{k,l>1} t_{kl}g_kg_l \le c_2(n-1)\det t^{1/(n-1)}.$$

Dispose of the integer g_1 in such a way that

$$\left|g_1 + \frac{y_{12}}{y_1}g_2 + \cdots + \frac{y_{1n}}{y_1}g_n\right| \le \frac{1}{2}.$$

Then we obtain from (4) and $y_1 = \mu(y)$

$$y_1 \le \tfrac{1}{4}y_1 + c_2(n-1)\det t^{1/(n-1)},$$

$$y_1^n \le (\tfrac{4}{3}c_2(n-1))^{n-1}y_1 \det t.$$

But $y_1 \det t = \det y$, so

$$y_1 \le (\tfrac{4}{3}c_2(n-1))^{(n-1)/n}\det y^{1/n} = c_2(n)\det y^{1/n}.$$

Remark

The demand for optimal constants $c_2(n)$ in Hermite's lemma and the corresponding quadratic forms is closely connected with the determination of the densest lattice packings of spheres. Optimal constants are only known for $1 \le n \le 8$.

Now we prove Minkowski's inequality by induction on n, where the case $n = 1$ is obvious. Let us consider any $y \in R_n$ and assume the assertion to be true for smaller values of n.

Case (i): The quotients of the diagonal elements satisfy

$$\frac{y_{\nu+1}}{y_\nu} \le \gamma_n \qquad (\nu = 1, \ldots, n-1)$$

with some constant γ_n depending only on n to be specified later. Then we have

$$\prod_{\nu=1}^{n} y_\nu \le \gamma_n^{n(n-1)/2} y_1^n.$$

Since $y \in R_n$ the quadratic form y has minimum y_1, and by Lemma 1

$$y_1^n \le c_2^n \det y.$$

From both inequalities we infer that Minkowski's inequality holds for any

$$c_1 \ge c_2^n \gamma_n^{n(n-1)/2}.$$

Case (ii): There exists an integer k, $1 \le k \le n-1$, such that

$$\frac{y_{k+1}}{y_k} > \gamma_n, \qquad \frac{y_{\nu+1}}{y_\nu} \le \gamma_n \qquad (\nu = k+1, \ldots, n-1). \tag{5}$$

Decompose y as in (3) with a k-rowed matrix p. Then by Minkowski's reduction conditions we obtain

$$y_{k+1} \le y[g] = p[g_1 + p^{-1}qg_2] + t[g_2], \qquad g = \begin{pmatrix} g_1 \\ g_2 \end{pmatrix}$$

for all $(k+1)$-admissible columns g. We choose g_2 by Lemma 1 such that

$$t[g_2] \le c_2(n-k) \det t^{1/(n-k)}, \qquad g_2 \text{ primitive},$$

and then g_1 such that the elements of $g_1 + p^{-1}qg_2$ lie between $-\frac{1}{2}$ and $+\frac{1}{2}$. Since p is reduced we infer from statements (i) and (ii) of Proposition 1 that

$$p[g_1 + p^{-1}qg_2] \le \tfrac{1}{8}k(k+1)y_k.$$

Combining the last three inequalities we get

$$y_{k+1} \le \tfrac{1}{8}k(k+1)y_k + c_2(n-k)\det t^{1/(n-k)}.$$

Now we contrast this estimate with the first inequality (5) to conclude that

$$y_{k+1}\left(1 - \frac{k(k+1)}{8\gamma_n}\right) \le c_2(n-k)\det t^{1/(n-k)}.$$

Then specify $\gamma_n := n(n-1)/4$, for instance, to obtain the crucial upper bound

$$y_{k+1} \le 2c_2(n-k)\det t^{1/(n-k)}. \tag{6}$$

In order to finish the proof we have only to collect our partial results on the diagonal elements of y. Since p is reduced the induction hypothesis yields

$$y_1 \cdots y_k \le c_1(k)\det p.$$

From the second group of inequalities (5) and (6) we obtain

$$\begin{aligned} y_{k+1}\cdots y_n &\le \gamma_n^{(n-k)(n-k-1)/2} y_{k+1}^{n-k} \\ &\le (2c_2(n-k))^{n-k}\gamma_n^{(n-k)(n-k-1)/2}\det t. \end{aligned}$$

Hence,

$$\prod_{\nu=1}^{n} y_\nu \le c_1(k)(2c_2(n-k))^{n-k}\gamma_n^{(n-k)(n-k-1)/2}\det y,$$

and Minkowski's inequality holds for any

$$c_1(n) \ge c_1(k)(2c_2(n-k))^{n-k}\gamma_n^{(n-k)(n-k-1)/2} \qquad (k = 1,\ldots,n-1).$$

Remark
Note that our proof allows the numerical computation of suitable constants for which Minkowski's inequality is valid.

Proposition 1 suggests considering the following more general regions.

Definition 2
For any positive integer n and positive real number t let

$$Q_n(t) = \{y \in P_n \mid y \text{ satisfies (i)–(iii)}\},$$

where

(i) $y_k < t y_{k+1} \quad (k = 1,\ldots,n-1)$,

(ii) $2|y_{kl}| < t y_k \quad (k < l)$,

(iii) $\prod_{\nu=1}^{n} y_\nu < c_1(n) t \det y$.

Here and henceforth $c_1(n)$ denotes any fixed choice of the constant in Minkowski's inequality.

These are open subsets of P_n which exhaust P_n for t tending to infinity; i.e. $Q_n(t) \subset Q_n(t')$ for $t < t'$, and any compact subset of P_n is contained in $Q_n(t)$ for sufficiently large t. We have $R_n \subset Q_n(t)$ for arbitrary $t > 1$ by Proposition 1.

Another kind of auxiliary regions that we need is defined by the Jacobian decompositions of positive matrices. It is well known from linear algebra, or can be derived immediately from (3), that any positive real matrix can be expressed uniquely as

$$y = d[v], \qquad d = \begin{pmatrix} d_1 & & 0 \\ & \ddots & \\ 0 & & d_n \end{pmatrix}, \qquad v = \begin{pmatrix} 1 & & * \\ & \ddots & \\ 0 & & 1 \end{pmatrix},$$

where d is a diagonal matrix and v is an upper triangular matrix with ones in the main diagonal. We call the d_ν ($\nu = 1,\ldots,n$) and the elements v_{kl} ($k < l$) of v 'Jacobian coordinates of y'.

Definition 3

For any positive integer n and positive real number t let

$$Q'_n(t) = \{y \in P_n \mid y = d[v] \textit{ satisfies (i), (ii)}\},$$

where

(i) $d_k < td_{k+1}$ $(k = 1,\ldots,n-1)$,
(ii) $|v_{kl}| < t$ $(k < l)$.

These again are open subsets of P_n which exhaust P_n as $t \to \infty$. Note that $Q'_n(t)$ is characterized as a subset of P_n by finitely many linear inequalities in the Jacobian coordinates.

The regions $Q_n(t)$ and $Q'_n(t)$ are equivalent in the following sense.

Proposition 2

Let n be a positive integer.

(i) *For any real number $t > 0$ there exists a $t_1 = t_1(n,t)$ depending only on n and t such that $Q_n(t) \subset Q'_n(t_1)$;*
(ii) *for any real number $s > 0$ there exists an $s_1 = s_1(n,s)$ depending only on n and s such that $Q'_n(s) \subset Q_n(s_1)$.*

Proof

From the Jacobian decomposition $y = d[v]$ we obtain for the diagonal elements

$$y_l = d_l\left(1 + \sum_{k=1}^{l-1} \frac{d_k}{d_l} v_{kl}^2\right) \qquad (l = 1,\ldots,n). \tag{7}$$

First we show that y_ν and d_ν ($\nu = 1,\ldots,n$) have the same order of magnitude. This is true under either assumption relevant in our proposition. Of course, (7) always implies

$$\frac{y_\nu}{d_\nu} \geq 1 \qquad (\nu = 1, \ldots, n).$$

Now assume $y \in Q_n(t)$. From condition (iii) in Definition 2 we infer

$$\prod_{\nu=1}^{n} \frac{y_\nu}{d_\nu} < c_1(n)t.$$

Therefore the quotients y_ν/d_ν are bounded from above and from below; we have

$$1 \leq \frac{y_\nu}{d_\nu} \leq t_2 \qquad (\nu = 1, \ldots, n),$$

where the constant t_2 depends only on n and t. If on the other hand $y \in Q'_n(s)$ we can deduce a similar result directly from (7) and Definition 3. Hence in both cases y_ν and d_ν are of the same order of magnitude.

We prove statement (i) by induction on n, where $n = 1$ is obvious. Assume the assertion to be true for $n - 1$ and consider any $y \in Q_n(t)$. In the Jacobian decomposition $y = d[v]$ we set

$$y = \begin{pmatrix} y^* & \eta \\ {}^t\eta & y_n \end{pmatrix}, \qquad d = \begin{pmatrix} d^* & 0 \\ 0 & d_n \end{pmatrix}, \qquad v = \begin{pmatrix} v^* & \omega \\ 0 & 1 \end{pmatrix},$$

where y^*, d^* and v^* are $(n-1)$-rowed. Then

$$y^* = d^*[v^*], \qquad y_n = d_n + d^*[\omega], \qquad \eta = {}^tv^* d^* \omega \tag{8}$$

and $y^* = d^*[v^*]$ is the Jacobian decomposition of y^*. Now $y \in Q_n(t)$ implies $y^* \in Q_{n-1}(t_3)$, where

$$t_3 = \max\left\{t, \frac{c_1(n)}{c_1(n-1)}t\right\}.$$

Indeed conditions (i) and (ii) of Definition 2 are trivially satisfied by y^* and condition (iii) is checked by

$$\frac{y_1 \cdots y_{n-1}}{\det y^*} < c_1(n) y_n^{-1} t \frac{\det y}{\det y^*} = c_1(n) t \frac{d_n}{y_n} \leq c_1(n)t.$$

By the induction hypothesis applied to $y^* \in Q_{n-1}(t_3)$ there exists a t_4 depending only on n and t such that

$$d_k < t_4 d_{k+1} \quad (k = 1, \ldots, n-2), \qquad |v_{kl}| < t_4 \quad (1 \leq k < l \leq n-1).$$

Since y_n and d_n have the same order of magnitude and $y_{n-1} < t y_n$, we may add $d_{n-1} < t_4 d_n$. So only the boundedness of ω is left. We obtain from (8)

$$\omega = d^{*-1}\, {}^tv^{*-1} \eta.$$

Now v^{*-1} is bounded as the inverse of a bounded triangular matrix. The boundedness of $d^{*-1}\eta$ follows from condition (ii) in Definition 2 and the fact that the y_ν and the d_ν have the same order of magnitude. All the bounds depend only on n and t.

The proof of the second statement is straightforward. Let y be any element of $Q'_n(s)$. Conditions (i) and (iii) in Definition 2 are satisfied for y as they are immediate consequences of y belonging to $Q'_n(s)$ and the boundedness of y_ν/d_ν $(\nu = 1,\ldots,n)$ from above and below by positive constants. Finally from the Jacobian decomposition $y = d[v]$ we obtain for the non-diagonal elements of y

$$|y_{kl}| = \left|\sum_{\nu=1}^{k} d_\nu v_{\nu k} v_{\nu l}\right| \leq s_2 d_k \leq s_2 y_k \qquad (k < l)$$

with $s_2 = s_2(n, s)$ again depending only on n and s. Thus the proposition is proved.

As an application we illustrate that to a certain extent a Minkowski reduced matrix may be replaced by its diagonalization. Denote by

$$y^{\mathrm{D}} = \begin{pmatrix} y_1 & & 0 \\ & \ddots & \\ 0 & & y_n \end{pmatrix}$$

the diagonal matrix made up of the diagonal elements $y_1, \ldots, y_n$ of the matrix y. This is called the diagonalization of y.

Lemma 2

Let F be one of the regions R_n, $Q_n(t)$ or $Q'_n(t)$. Then there exists a positive number γ depending only on n and t such that

$$\gamma^{-1} y < y^{\mathrm{D}} < \gamma y$$

for arbitrary $y \in F$.

Proof

Since $R_n \subset Q_n(t)$ for $t > 1$ and $Q_n(t) \subset Q'_n(t')$ for appropriate t' by Proposition 2, it is sufficient to consider $Q'_n(t)$. Let $y = d[v]$ be the Jacobian decomposition of any $y \in Q'_n(t)$ and

$$x = d^{1/2} v d^{-1/2}.$$

Then x is bounded by the reduction conditions in Definition 3. Since x is a triangular matrix, x^{-1} is bounded too. Hence

$$y = {}^t x x[d^{1/2}]$$

has the same order of magnitude as d, i.e. there exists a positive number $\gamma_1 = \gamma_1(n, t)$ such that

$$\gamma_1^{-1} d < y < \gamma_1 d.$$

On the other hand, we have already seen in the proof of Proposition 2 that d is of the same order of magnitude as y^{D},

$$\gamma_2^{-1} d < y^{\mathrm{D}} < \gamma_2 d.$$

Hence the lemma holds for $\gamma = \gamma_1 \gamma_2$.

To summarize our results, we have introduced three regions R_n, $Q_n(t)$ and $Q'_n(t)$. The last two of them are each contained in the other as in Proposition 2. R_n is contained in $Q_n(t)$, respectively $Q'_n(t)$, for sufficiently large t. The formal description as a subset of P_n varies in quality: there are infinitely many linear inequalities for R_n, finitely many algebraic inequalities for $Q_n(t)$, finitely many linear inequalities in the Jacobian coordinates for $Q'_n(t)$. We know that the images of each of those three regions (t sufficiently large) under the action of U_n cover all of P_n. We now turn to the question of whether there are only finitely many overlappings. It is because of this problem that we have introduced the auxiliary regions $Q_n(t)$ and $Q'_n(t)$. Moreover we will be able to analyze the geometrical structure of R_n on this occasion. We now state the most important theorem in reduction theory.

Theorem 1

Let $y, y^* \in Q'_n(t)$, $y^* = y[g]$, and g be any n-rowed integral matrix such that $|\det g| \leq t$. Then the elements of g are bounded by some constant depending only on n and t.

Corollary (i)

Let F be any of the regions R_n, $Q_n(t)$ or $Q'_n(t)$. Then the set of unimodular matrices

$$\{u \in U_n \mid F \cap u(F) \neq \emptyset\}$$

is finite.

This corollary is an immediate consequence of the theorem and the fact that R_n and $Q_n(t)$ may be enclosed in $Q'_n(t')$ for appropriate t'. We define a fundamental set for a group action as a subset of the representation space such that the images of the subset cover the representation space and the finiteness condition of the corollary is satisfied. In our case R_n, $Q_n(t)$ and $Q'_n(t)$ for sufficiently large t are fundamental sets for the unimodular group.

Any fundamental set contains at least one and at most finitely many representatives of each orbit. The number of representatives is bounded independently of the individual orbit. Finally we use the fact that each compact subset of P_n can be enclosed in $Q'_n(t)$ for an appropriate choice of t. So we may state

Corollary (ii)

R_n, $Q_n(t)$ and $Q'_n(t)$, for sufficiently large t, are fundamental sets for the group action of the unimodular group U_n on the space P_n of positive definite quadratic forms. Each compact subset of P_n is covered by at most finitely many images of these fundamental sets.

Proof

The theorem will be proved by induction on n, where again the case $n = 1$ is obvious. We distinguish two cases for $n > 1$.

Case (i): Assume the existence of a number k, $1 \leq k < n$, such that g splits into

$$g = \begin{pmatrix} g_1 & g_{12} \\ 0 & g_2 \end{pmatrix},$$

where g_1 is k-rowed. During the course of this proof we use the corresponding notation for other matrices decomposed in a similar way. Let

$$y^* = d^*[v^*], \qquad y = d[v]$$

be the Jacobian decompositions of y, y^* such that $d^* = d[vgv^{*-1}]$. This last condition may be reformulated as the orthogonality of

$$d^{1/2}vgv^{*-1}d^{*-1/2}.$$

Now this matrix has the same shape as g, i.e. the block in the lower left corner is zero. Hence orthogonality means that the block in the upper right corner vanishes and the blocks along the diagonal are orthogonal. So we obtain

$$d_1^{1/2}v_1g_1v_1^{*-1}d_1^{*-1/2} \text{ orthogonal} \quad \Leftrightarrow \quad d_1[v_1g_1] = d_1^*[v_1^*],$$

$$d_2^{1/2}v_2g_2v_2^{*-1}d_2^{*-1/2} \text{ orthogonal} \quad \Leftrightarrow \quad d_2[v_2g_2] = d_2^*[v_2^*],$$

$$v_1g_{12} + v_{12}g_2 = v_1g_1v_1^{*-1}v_{12}^*.$$

Since $|\det g_1|$, $|\det g_2| \leq t$ we infer the boundedness of g_1 and g_2 from the first two equations and the induction hypothesis. The boundedness of g_{12} can be seen from the last equation. All the bounds depend only on n and t.

Case (ii): To each k, $1 \leq k < n$, there exists an $r > k$ and an $s \leq k$ such that $g_{rs} \neq 0$. Since y, $y^* \in Q_n'(t)$ the matrices y, y^* have the same order of magnitude as d, d^*, respectively; consequently we have

$$t_1^{-1} d[g] \leq y[g] = y^* < t_1 d^*$$

for a certain t_1 depending only on n and t. In particular,

$$\sum_{\nu=1}^{n} d_\nu g_{\nu\mu}^2 \leq t_1^2 d_\mu^* \qquad (\mu = 1, \ldots, n). \tag{9}$$

But g_{rs} is integral and different from zero, so

$$d_r \leq t_1^2 d_s^*.$$

The sequences $d_1, \ldots, d_n$ and $d_1^*, \ldots, d_n^*$ are monotonically increasing up to the factor t because y, y^* belong to $Q_n'(t)$. Therefore we obtain

$$d_{k+1} \leq t_2 d_k^* \qquad (k = 1, \ldots, n-1)$$

with an appropriate constant $t_2 = t_2(n, t)$. Now apply the same argument to the equation

$$\det g^2\, y = y^*[\det g\, g^{-1}].$$

Note that $\det g\, g^{-1}$ is again integral, with determinant less than or equal to t^{n-1} and of the relevant form for case (ii). Hence we have

$$d_{k+1}^* \leq t_3 d_k \qquad (k = 1, \ldots, n-1)$$

with another constant $t_3 = t_3(n, t)$. From the last two inequalities and the natural ordering (up to the factor t) of $d_1, \ldots, d_n$ and $d_1^*, \ldots, d_n^*$ we deduce the boundedness of all the quotients

$$\frac{d_k^*}{d_l} \qquad (k, l = 1, \ldots, n).$$

Finally, return to (9) in order to see the boundedness of $g_{\nu\mu}$ $(\nu, \mu = 1, \ldots, n)$, and conclude the proof of the theorem.

In the remainder of this section we will study Minkowski's reduced domain R_n more closely. We are interested in the geometrical properties of R_n and the structure of the covering of P_n by R_n and its images under the unimodular group. First omit all the reduction conditions in Definition 1, which are satisfied identically in $y \in P_n$. Those which are left are exactly

$$y[g] \geq y_k \quad \text{for all } k\text{-admissible } g \neq \pm e_k \quad (k = 1, \ldots, n), \tag{10'}$$

$$y_{k,k+1} \geq 0 \quad (k = 1, \ldots, n-1), \tag{10''}$$

called proper reduction conditions. Thus R_n as a subset of P_n appears as the intersection of countably many half-spaces, defined by the individual linear inequalities (10). First we show that there exist points $y \in R_n$ such that all these inequalities hold with the $>$ sign, henceforth called strict inequalities. Indeed consider the countably many hyperplanes defined by (10) with the sign of equality and their countably many images under U_n. Choose a point $y \in P_n$ which does not lie on any of those hyperplanes. Determine $u \in U_n$ such that $y[u] = y^*$ is reduced in the sense of Minkowski. Then y^* fulfils our requirements. According to the next proposition such a point is an interior point of R_n.

We consider R_n as a subset of P_n from a topological point of view. Obviously R_n is closed in P_n.

Proposition 3

The interior and the boundary of Minkowski's reduced domain R_n in P_n are the following subsets:

$$\mathring{R}_n = \{y \in P_n \mid y \text{ satisfies the strict inequalities } (10)\},$$

$$\partial R_n = \{y \in P_n \mid y \text{ satisfies } (10), \text{ at least once with the sign of equality}\}.$$

Proof

It is sufficient to show that the sets on the right are contained in $\mathring{R}_n$, ∂R_n, respectively. Let $y \in P_n$ fulfil the strict inequalities (10). If λ is the lowest eigenvalue of y, choose $0 < \varepsilon < \lambda$ and a neighborhood V of y such that

$$y^* - y > -\varepsilon \mathbf{1}, \qquad |y_k^* - y_k| < \varepsilon, \qquad y_{k,k+1}^* > 0$$

for all $y^* \in V$. Then

$$\begin{aligned} y^*[g] - y_k^* &= y[g] + (y^* - y)[g] + y_k - y_k^* - y_k \\ &\geq (\lambda - \varepsilon)\,{}^t gg - \varepsilon - y_k. \end{aligned}$$

The expression on the right does not depend on y^* and is positive for all but finitely many integral columns g. Hence the inequalities (10), with the exception of at most finitely many k-admissible g, hold for arbitrary $y^* \in V$. Diminishing V afterwards, the exceptions may be included such that V is contained in R_n. Now let $y \in P_n$ satisfy (10) and at least one particular inequality with the sign of equality. Since we have omitted the identities among the reduction conditions, there are points of P_n arbitrarily close to y for which this particular inequality fails. Hence y is a boundary point.

Next we are able to deduce a better understanding of the covering of P_n by R_n and its images under the unimodular group. Since $\pm u$ induce the same automorphism of P_n, let us consider more precisely the action of

$U_n/\{\pm\mathbf{1}\}$ on P_n. The next proposition states that overlappings may occur only on the boundary.

Proposition 4

Let y, $y^* \in R_n$, $y^* = y[u]$ and $\pm\mathbf{1} \neq u \in U_n$. Then y and y^* belong to the boundary of R_n.

Proof

If u is a diagonal matrix we may assume for the columns $u_1 = e_1, \ldots, u_k = e_k$, but $u_{k+1} = -e_{k+1}$. From $y^* = y[u]$ we obtain

$$y^*_{k,k+1} = -y_{k,k+1}.$$

On the other hand, both quantities are non-negative by the reduction conditions. Hence,

$$y^*_{k,k+1} = y_{k,k+1} = 0$$

and y, $y^* \in \partial R_n$ by Proposition 3. If u is non-diagonal, let $u_1 = \pm e_1, \ldots, u_{k-1} = \pm e_{k-1}$ but $u_k \neq \pm e_k$. Then u_k is k-admissible, and by Minkowski's reduction conditions

$$y^*_k = y[u_k] \geq y_k.$$

Interchanging y and y^* and replacing u by u^{-1} we obtain $y_k \geq y^*_k$. Hence,

$$y[u_k] = y_k, \qquad u_k \neq \pm e_k$$

and $y \in \partial R_n$, likewise $y^* \in \partial R_n$, by Proposition 3.

Finally we are able to analyze the geometrical structure of R_n completely. Consider the set

$$V_n = \{u \in U_n | R_n \cap u(R_n) \neq \emptyset\}.$$

It is a finite set by Corollary (i) of Theorem 1. The rows of the matrices in V_n form another finite set, denoted by V_n^*.

Proposition 5

R_n as a subset of P_n is the intersection of the finitely many half-spaces defined by

$$y[g] \geq y_k, \quad g \text{ } k\text{-admissible}, g \neq \pm e_k, {}^t g \in V_n^* \quad (k = 1, \ldots, n), \tag{11'}$$

$$y_{k,k+1} \geq 0 \quad (k = 1, \ldots, n-1). \tag{11''}$$

Corollary

Minkowski's reduced domain R_n is a convex pyramid bounded by finitely many hyperplanes and with vertex at the origin.

Proof

Let y be a boundary point of R_n. By Proposition 3, at least one of the infinitely many reduction conditions (10) is fulfilled with the sign of equality, for instance

$$y[g] = y_k, \qquad g\ k\text{-admissible}, \qquad g \neq \pm e_k.$$

From the construction of reduced forms by minimizing conditions at the beginning of this section we immediately deduce that

$$(\pm e_1, \ldots, \pm e_{k-1}, g)$$

is complementary to a unimodular matrix u, such that $y[u] \in R_n$. Hence, ${}^t g$ belongs to V_n^*. Note that the conditions (10″) appear again as (11″). So we may state that for any $y \in \partial R_n$ at least one of the finitely many conditions (11) holds with the sign of equality. In order to prove the proposition assume there is a point $y \in P_n$ satisfying (11) and not belonging to R_n. Connect y with an interior point y^* of R_n by the line segment

$$\lambda y + (1 - \lambda) y^* \qquad (0 < \lambda < 1).$$

Then, on the one hand, the strict inequalities (11) hold for each point of this segment, since (11) are satisfied for y^* in the strict sense and for y in the usual sense. Furthermore, the reduction conditions are linear homogeneous inequalities. On the other hand, there must be a boundary point of R_n on this line segment, and at that point one of the conditions (11) holds with the sign of equality as we have already seen. This contradiction finishes the proof.

So we have finally proved that the infinitely many original reduction conditions of Minkowski are consequences of a finite subset. Summarizing our results we state the following

Theorem 2

Minkowski's reduced domain R_n as a subset of P_n is a convex pyramid bounded by finitely many hyperplanes. The images of R_n under the group $U_n/\{\pm 1\}$ cover P_n without gaps and essential overlappings. R_n has only finitely many neighbors and each compact subset of P_n is covered by finitely many images of R_n.

Among many important applications of Minkowski's reduction theory we mention only one, which is concerned with group-theoretical consequences. It was proved by M. Gerstenhaber [25] and H. Behr [6] under conditions of wide generality that one may deduce a finite presentation of the underlying group from the properties in the theorem. Generators are the transforms which map R_n into its neighbors, called local generators; defining relations are the local relations – the possible relations of the form $u_1 \cdot u_2 = u_3$ between the local generators. For generalizations of Minkowski's reduction theory see A. Borel [11].

3 Fundamental sets of Siegel's modular group

In this section we return to the action of the symplectic group $Sp(n, \mathbb{R})$ on Siegel's half-space H_n as in Proposition 1.1. If G is any subgroup of $Sp(n, \mathbb{R})$, we denote by $\hat{G}$ the corresponding group of induced automorphisms

$$z \mapsto m\langle z\rangle \qquad (m \in G)$$

of H_n. The group G acts discontinuously on H_n – or, synonymously, $\hat{G}$ is discontinuous – if the family

$$\{m\langle z\rangle | m \in G\}$$

has no accumulation point in H_n, no matter what $z \in H_n$ may be. Since H_n is the union of countably many compact subsets, each discontinuous group $\hat{G}$ is enumerable. On the other hand, G as a subgroup of the topological group $Sp(n, \mathbb{R})$ is called discrete if there exists a neighborhood U of the unit element $\mathbf{1}$ such that no other element of G is contained in U. Obviously this is equivalent to the fact that no sequence (g_ν) of mutually distinct elements of G converges in G (or in the space of all $2n \times 2n$ real matrices). In our case we have

Proposition 1

Let G be any subgroup of $Sp(n, \mathbb{R})$. Then G acts discontinuously on H_n if and only if G is discrete.

Proof

It is obvious that discontinuity implies discreteness. Conversely, let the action of G be non-discontinuous; then there exists a point $z \in H_n$ and a sequence of mutually distinct elements m_ν in G such that

$$z_\nu^* = m_\nu\langle z\rangle \qquad (\nu = 1, 2, \ldots)$$

converges to z^* in H_n. For the imaginary parts we obtain

$$y_\nu^{*-1} = y^{-1}\{{}^t(c_\nu z + d_\nu)\}$$

by (1.6). Since y_ν^{*-1} is bounded with respect to ν and y^{-1} independent of ν, we can infer the boundedness of $c_\nu z + d_\nu$. Decomposing these matrices into their real and imaginary parts, c_ν and d_ν turn out to be bounded as well.

Apply the same argument to

$$a_\nu z + b_\nu = z_\nu^*(c_\nu z + d_\nu)$$

to get a similar result for a_ν and b_ν. Hence the sequence (m_ν) is bounded and consequently there exists a convergent subsequence, proving G to be non-discrete.

The most important example of a discrete subgroup of $Sp(n, \mathbb{R})$ is

$$\Gamma_n := Sp(n, \mathbb{Z}),$$

called Siegel's modular group. We consider the action of Γ_n on H_n induced by the former action of the symplectic group. This section is devoted to the determination of different fundamental sets of Γ_n which are useful in studying the corresponding automorphic forms and functions. For $n = 1$ we obtain the well-known elliptic modular group; keeping this special case in mind, the reader will certainly note that we are not considering subgroups, in particular congruence subgroups of the modular group. Indeed we shall restrict ourselves to the full modular group in order to keep the whole theory as transparent as possible; subgroups will be mentioned only sporadically.

To construct fundamental sets for Siegel's modular group we introduce three subgroups of Γ_n:

$$\left\{\begin{pmatrix} u & 0 \\ 0 & {}^t u^{-1} \end{pmatrix} \,\middle|\, u \text{ unimodular}\right\} \simeq U_n,$$

$$\left\{\begin{pmatrix} 1 & s \\ 0 & 1 \end{pmatrix} \,\middle|\, s \text{ symmetric, integral entries}\right\}, \tag{2}$$

$$\left\{\begin{pmatrix} u & s\,{}^t u^{-1} \\ 0 & {}^t u^{-1} \end{pmatrix} \,\middle|\, u \text{ unimodular; } s \text{ symmetric, integral entries}\right\}. \tag{3}$$

The first one is canonically isomorphic to the unimodular group; the second one is the group of translations, a free Abelian group of rank $n(n + 1)/2$; the third consists of all elements of Γ_n for which the lower left block vanishes, sometimes called integral modular substitutions.

First consider the 'second rows' (c, d) of modular matrices $\begin{pmatrix} a & b \\ c & d \end{pmatrix}$. Of course these lower blocks are matrices of n rows and $2n$ columns. Looking at the left cosets of Γ_n modulo the first subgroup mentioned above, we realize that one may multiply any second row (c, d) by an arbitrary $u \in U_n$ from the left, getting another second row of a modular substitution. Two second rows are called associated if they differ only by a factor $u \in U_n$ on the left.

Lemma 1

For fixed $z \in H_n$ and any real number $\lambda > 0$ there exist only finitely many classes of associated second rows (c, d) of modular substitutions such that

$$|\det(cz + d)| < \lambda.$$

Proof

Consider any $m \in \Gamma_n$, the second row (c, d) of which satisfies the inequality above. Then we infer from (1.6) for the imaginary part y^* of $m\langle z\rangle$

$$y^{*-1} = y^{-1}\{{}^t(cz + d)\}.$$

Multiply c and d by an appropriate $u \in U_n$ such that y^{*-1} becomes reduced in the sense of Minkowski. Denote the diagonal elements of y^{*-1} by $y_1^*, \ldots, y_n^*$ and the rows of c and d by $c_1, \ldots, c_n$, respectively $d_1, \ldots, d_n$. Then we obtain

$$y_l^* = y^{-1}[{}^t(c_l x + d_l)] + y[{}^t c_l] \qquad (l = 1, \ldots, n). \tag{4}$$

Since $y > 0$ and c_l, d_l are integral and not both zero, the y_l^* are bounded from below by a positive constant independent of (c, d). On the other hand, we have

$$\det y^{*-1} \leq \lambda^2 \det y^{-1},$$

and we may replace y^{*-1} in this inequality up to a constant by its diagonalization because of Lemma 2.2. Hence the product of all the y_l^* is bounded from above. So finally we obtain that each y_l^* is bounded from above and below by positive constants independent of (c, d). Then, looking once again at (4), only finitely many choices for c and d are left.

Analogous to the one-variable case, we call $\det y$ the 'height' of the point z in H_n. From the transformation formula (1.6) for the imaginary parts and the lemma above we realize that each Γ_n-orbit contains points of maximal height. Such points are characterized by the conditions

$$|\det(cz + d)| \geq 1$$

for all $m \in \Gamma_n$. Furthermore, the height remains unchanged if one applies an integral modular substitution. So we can dispose of u and s in the subgroups (1) and (2) such that y becomes reduced in the sense of Minkowski and the real part x reduced modulo 1. Hence each Γ_n-orbit contains points of the following set F_n.

Definition 1

Let n be any positive integer. Siegel's fundamental domain is the subset

$$F_n = \{z \in H_n \mid z \text{ satisfies (i)–(iii)}\}$$

of H_n, where

(i) $|\det(cz + d)| \geq 1$ *for all $m \in \Gamma_n$,*
(ii) $y \in R_n$,
(iii) $|x_{kl}| \leq \frac{1}{2}$ $(k \leq l;\ k, l = 1, \ldots, n)$.

Here R_n denotes Minkowski's reduced domain as in §2. For $n = 1$ we obtain the well-known standard fundamental domain in the upper half-plane with the cusp at infinity. In the general case this fundamental domain had already been introduced by C.L. Siegel in his early papers [62] and [64].

Another elementary geometrical concept which we borrow from the one-variable case is that of a 'vertical strip of positive height d' defined by

$$V_n(d) = \{z \in H_n \mid \sigma(x^2) \leq d^{-1}, y \geq d\mathbf{1}\}.$$

Here d is any positive number and σ denotes the trace. Clearly these sets exhaust H_n for d tending to zero. The symplectic volume of a vertical strip is finite. Indeed, the symplectic volume element in §1 may be rewritten as

$$\mathrm{d}v = \mathrm{d}x\,\mathrm{d}y^{-1},$$

and x and y^{-1} are bounded in any vertical strip by definition. A simple observation yields

Lemma 2

There exists a positive number d such that Siegel's fundamental domain F_n is contained in $V_n(d)$.

Proof

The real part x is bounded for any $z \in F_n$ by condition (iii). Concerning the imaginary part we have to show that $y \geq d\mathbf{1}$ for an appropriate positive number d and all $z = x + iy \in F_n$. Since y is reduced we may replace y by its diagonalization in accordance with Lemma 2.2. Furthermore the diagonal elements are increasing by Proposition 2.1. So it is sufficient to prove that the first diagonal element y_1 of y is bounded from below by a positive constant. But this can be deduced immediately from condition (i) for $m = \begin{pmatrix} a & b \\ c & d \end{pmatrix}$, where

$$a = \begin{pmatrix} 0 & 0 \\ 0 & \mathbf{1} \end{pmatrix}, \quad b = \begin{pmatrix} 1 & 0 \\ 0 & 0 \end{pmatrix}, \quad c = \begin{pmatrix} -1 & 0 \\ 0 & 0 \end{pmatrix}, \quad d = \begin{pmatrix} 0 & 0 \\ 0 & \mathbf{1} \end{pmatrix}.$$

These decompositions are of type $(1, n-1)$. Condition (i) says that $|z_1| \geq 1$, which together with $|x_1| \leq \frac{1}{2}$ implies that $y_1 \geq \sqrt{3}/2$.

Corollary

Siegel's fundamental domain is closed in the space of all symmetric complex matrices.

Now we discuss the question of whether F_n is a fundamental set in the sense introduced in §2. For this purpose it is covenient to enlarge F_n in a way similar to the procedure used in Minkowski's reduction theory.

Definition 2

For any positive integer n and positive real number t let

$$L_n(t) = \left\{ z \in H_n \,\middle|\, |x_{kl}| < t,\ y \in Q_n'(t),\ y_1 > \frac{1}{t} \right\}.$$

We refer to Definition 2.3 for the meaning of $Q_n'(t)$. These are open subsets of H_n, which exhaust H_n for $t \to \infty$. Furthermore, $F_n \subset L_n(t)$ for sufficiently large t, since then $R_n \subset Q_n'(t)$, and y_1 is bounded from below for any $z \in F_n$ by Lemma 2. These sets $L_n(t)$ can still be enclosed in vertical strips of positive height and have therefore finite symplectic volume.

Theorem 1

The set

$$\{m \in \Gamma_n \mid L_n(t) \cap m\langle L_n(t)\rangle \neq \emptyset\}$$

is finite for any given real number $t > 0$.

Corollary

The sets F_n and $L_n(t)$ for sufficiently large t are fundamental sets for the group action of Γ_n on H_n. Each compact subset of H_n is covered by at most finitely many images of these fundamental sets under Γ_n.

Remark

Note that vertical strips of positive height do not satisfy the finiteness condition of Theorem 1.

Proof

The idea is to transform the problem in question into the space P_{2n} of positive definite quadratic forms of degree $2n$ over $\mathbb{R}$ and to apply our former results on reduction theory. Consider

$$s = \begin{pmatrix} y^{-1} & 0 \\ 0 & y \end{pmatrix} \left[\begin{pmatrix} 1 & -x \\ 0 & 1 \end{pmatrix} \right] \tag{5}$$

for arbitrary $z = x + iy \in H_n$. Then the assignment $z \mapsto s$ defines a bijective map from H_n onto the space S of all positive definite $2n$-rowed symplectic matrices,

$$H_n \to S = \{s \in Sp(n, \mathbb{R}) \mid s > 0\}.$$

Indeed, s defined by (5) is positive since y is positive, and symplectic because each factor is symplectic. To show surjectivity, decompose any $s \in S$ by completion of squares (cf. (2.3)) into

$$s = \begin{pmatrix} p & q \\ {}^t q & r \end{pmatrix} = \begin{pmatrix} p & 0 \\ 0 & t \end{pmatrix} \left[\begin{pmatrix} 1 & p^{-1}q \\ 0 & 1 \end{pmatrix} \right].$$

The symplecticity of s implies the symmetry of $p^{-1}q$ and $t = p^{-1}$; furthermore p is positive. Hence $z = -p^{-1}q + ip^{-1}$ belongs to H_n and is mapped onto s. Since (5) can be resolved uniquely with respect to x and y the map is injective. The correspondence between H_n and S has remarkable properties. First it is easy to check how the operation of $Sp(n, \mathbb{R})$ on H_n transforms to S. The following equivalence holds:

$$z^* = m\langle z \rangle \quad \Leftrightarrow \quad s^* = s[m^{-1}] \qquad (m \in Sp(n, \mathbb{R})). \tag{6}$$

Then our mapping turns out to be compatible with Jacobian decompositions in the following sense. If $z = x + iy$ and s are related by (5) and if $y = d[v]$ is the Jacobian decomposition of y, then

$$s\left[\begin{pmatrix} w & 0 \\ 0 & 1 \end{pmatrix}\right] = \begin{pmatrix} d^{-1}[w] & 0 \\ 0 & d \end{pmatrix} \left[\begin{pmatrix} w\,{}^t v^{-1} w & -w\,{}^t v^{-1} x \\ 0 & v \end{pmatrix}\right] \tag{7}$$

is again a Jacobian decomposition. Here w denotes the $n \times n$ matrix $\begin{pmatrix} 0 & & 1 \\ & \cdot^{\cdot^{\cdot}} & \\ 1 & & 0 \end{pmatrix}$. Now if z belongs to $L_n(t)$ then

$$s\left[\begin{pmatrix} w & 0 \\ 0 & 1 \end{pmatrix}\right] \in Q'_{2n}(\tau) \tag{8}$$

for an appropriate τ depending only on n and t. This is trivial as far as the triangular matrix in (7) is concerned, since v and x are bounded by t. The diagonal matrix in (7) has as diagonal elements $d_n^{-1}, \ldots, d_1^{-1}, d_1, \ldots, d_n$. Since $y \in Q'_n(t)$ this sequence increases up to the factor t everywhere except for the nth step. But there we have $d_1^{-1} < t^2 d_1$, since $d_1 = y_1 > 1/t$. Now using (6) and (8) the assertion follows from the corresponding Corollary (i) of Theorem 2.1 in Minkowski's reduction theory.

We are interested in studying in greater detail the geometrical properties of F_n and the structure of the covering of H_n by F_n and its images under the modular group. Since $\pm m$ induce the same automorphism of H_n, we consider more precisely the action of $\Gamma_n/\{\pm 1\}$ on H_n. The next proposition states that overlappings can occur only on the boundary.

Proposition 2

Let $z, z^* \in F_n$, $z^* = m\langle z\rangle$ and $\pm 1 \neq m \in \Gamma_n$. Then z and z^* are boundary points of F_n.

Proof

It is sufficient to consider z. Since z and z^* belong to the same orbit and both have maximal height we infer

$$|\det(cz + d)| = 1.$$

Hence either z is a boundary point of F_n or $c = 0$. In the latter case the imaginary parts y and y^* satisfy $y^* = y[u]$ with $u \in U_n$. If $u \neq \pm 1$, the point y belongs to the boundary of R_n in P_n by Proposition 2.4, hence z belongs to the boundary of F_n. In the remaining case m is a non-trivial translation, and condition (iii) of Definition 1 guarantees that z is a boundary point of F_n.

We now want to show that the infinitely many conditions in Definition 1 may be replaced by a finite subset. Introduce the set

$$V_n = \{m \in \Gamma_n \mid m \neq \pm 1, F_n \cap m\langle F_n\rangle \neq \varnothing\},$$

which is finite by Theorem 1, and rewrite the finitely many Minkowski reduction conditions (2.11) as linear inequalities $\varphi_\nu(y) \geq 0$ $(\nu = 1, \ldots, k)$. Then we may state

Proposition 3

Siegel's fundamental domain F_n consists of all $z \in H_n$ which satisfy the following finitely many inequalities:

(i) $|\det(cz + d)| \geq 1$ *for all* $m \in V_n$, $c \neq 0$,
(ii) $\varphi_\nu(y) \geq 0$ $(\nu = 1, \ldots, k)$,
(iii) $|x_{kl}| \leq \frac{1}{2}$ $(k \leq l;\ k, l = 1, \ldots, n)$.

Corollary

Siegel's fundamental domain is bounded by finitely many algebraic surfaces in H_n, which are described by the individual conditions above with the sign of equality.

For the proof of this proposition we need

Lemma 3

For fixed $m \in Sp(n, \mathbb{R})$ with $c \neq 0$ and fixed $z \in H_n$ set $z_\lambda = x + i\lambda y$. Then the inequality

$$|\det(cz_\lambda + d)| > |\det(cz_\mu + d)|$$

holds for all real $\lambda > \mu > 0$.

Proof

Consider the polynomial

$$\varphi(\lambda) = \det(cz_\lambda + d)$$

in the complex variable λ. Since

$$\overline{\varphi(\lambda)} = \varphi(-\bar{\lambda})$$

and $\varphi(\lambda) \neq 0$ for $\operatorname{Re}\lambda > 0$, the roots of φ are purely imaginary numbers. Therefore $|\varphi(\lambda)|^2$, considered for real λ, turns out to be a polynomial in λ^2 with non-negative real coefficients. This polynomial is non-constant since $c \neq 0$; hence $|\varphi(\lambda)|$ is strictly increasing for $\lambda > 0$.

In order to prove Proposition 3, let $z \in H_n$ satisfy the finitely many conditions (i)–(iii). Assume an $m \in \Gamma_n$ such that

$$|\det(cz + d)| \leq 1,$$

the second row of m non-associated to the second row of any element of V_n and $c \neq 0$. By Lemma 1 there are only finitely many classes of such second rows (c, d). Hence, using Lemma 3, we may choose $\lambda > 0$ such that $z_\lambda \in F_n$ and at least once the equality

$$|\det(cz_\lambda + d)| = 1$$

holds. Then both z_λ and $m\langle z_\lambda \rangle$ have maximal height. Transform $m\langle z_\lambda \rangle$ into F_n by an integral modular substitution m^* (cf. (3)). Then m^*m belongs to V_n, contradicting the assumptions on m. Thus we have proved

$$|\det(cz + d)| > 1$$

for all (c, d) not associated to second rows of elements of V_n with $c \neq 0$.

Summarizing our results we obtain

Theorem 2

Siegel's fundamental domain F_n is a closed subset of H_n bounded by finitely many algebraic surfaces. The images of F_n under the group $\Gamma_n/\{\pm 1\}$ cover H_n without gaps and essential overlappings. F_n has only finitely many neighbors, and each compact subset of H_n is covered by finitely many images of F_n.

In §2 we mentioned a result of M. Gerstenhaber and H. Behr concerning finite presentations of groups. One can apply this method to Siegel's modular group using the covering of H_n by F_n and its images as described above. Therefore we may deduce from our theorem the existence of finitely many generators and finitely many defining relations for Γ_n. On the other hand, we cannot expect an elegant presentation by this procedure. To illustrate the situation we mention a result of E. Gottschling [28] stating that Siegel's fundamental domain of degree two is already bounded by as many as 28 algebraic surfaces.

Another remark is concerned with the symplectic volume of F_n. Certainly F_n is not compact, since, for instance, $i\lambda\mathbf{1}$ belongs to F_n for arbitrary $\lambda \geq 1$. Nevertheless the symplectic volume of F_n is finite, since F_n may be included in a vertical strip of positive height by Lemma 2. Now since we know F_n to be bounded by finitely many algebraic surfaces, we are certain that F_n is measurable even in the Riemann sense. The value of the symplectic volume was computed by C.L. Siegel [64]. His result

$$\mathrm{vol}(F_n) = 2\pi^{-n(n+1)/2} \prod_{k=1}^{n} (k-1)!\zeta(2k)$$

yields an interesting connection with Riemann's zeta-function.

The construction of Siegel's fundamental domain depends essentially on the arithmetic of quadratic forms. Of course there are more general methods of finding fundamental domains if one refrains from some of the properties mentioned above. For instance, C.L. Siegel considered a different method in [64] which works for arbitrary subgroups of $Sp(n, \mathbb{R})$ operating discontinuously on H_n and which is based on the following idea of H. Poincaré. Choose an arbitrary non-fixed point z^* and select from each orbit a representative which has minimal symplectic distance from z^*. This particular fundamental domain has never been used in connection with Siegel's modular group.

Another method, which also goes back to Poincaré, is concerned with arbitrary discontinuous groups acting holomorphically on a bounded domain. We consider the procedure in detail for a subgroup Γ of $Sp(n, \mathbb{R})$ operating discontinuously on H_n. Although we formulate our results in H_n, we really work with the bounded model D_n introduced in Definition 1.3. The transition from H_n to D_n is performed by Cayley's transformation

$$w = l\langle z\rangle = (z - i\mathbf{1})(z + i\mathbf{1})^{-1}$$

as in Proposition 1.2. Then Γ has to be replaced by the subgroup $\tilde{\Gamma} = l\Gamma l^{-1}$ of Φ_n. The Jacobian of Cayley's transformation turns out to be

$$\det\left(\frac{\partial l\langle z\rangle}{\partial z}\right) = (2i)^{n(n+1)/2} \det(z + i\mathbf{1})^{-n-1},$$

and for $m \in Sp(n, \mathbb{R})$ we have

$$\det\left(\frac{\partial m\langle z\rangle}{\partial z}\right) = \det(cz + d)^{-n-1}.$$

Using the cocycle relation we are able to compute the Jacobian of $\tilde{m} = lml^{-1} \in \Phi_n$ in terms of H_n; after taking $(n + 1)$th roots we obtain

$$j_m(z) := \det\left(\frac{\partial \tilde{m}\langle w\rangle}{\partial w}\right)^{1/(n+1)} = \frac{\det(z + i\mathbf{1})}{\det(m\langle z\rangle + i\mathbf{1})\det(cz + d)}.$$

On the other hand we have

$$\det\left(\frac{\partial \tilde{m}\langle w\rangle}{\partial w}\right) = \det(\bar{\tilde{b}}w + \bar{\tilde{a}})^{-n-1}, \qquad \tilde{m} = \begin{pmatrix} \tilde{a} & \tilde{b} \\ \bar{\tilde{b}} & \bar{\tilde{a}} \end{pmatrix} \in \Phi_n.$$

In the next proposition we summarize some of the basic properties of these Jacobians which are essential for the construction of fundamental domains.

Proposition 4

The quantities $j_m(z)$ satisfy

(i) $j_{m_1m_2}(z) = j_{m_1}(m_2\langle z\rangle)j_{m_2}(z) \quad (m_1, m_2 \in Sp(n, \mathbb{R}))$,

(ii) $|j_m(z)| = 1$ *identically in* $z \in H_n$, *if and only if* m *is orthogonal, or equivalently iff* $\tilde{m}$ *is unitary,*

(iii) $\sum_{m\in\Gamma} |j_m(z)|^{2(n+1)}$ *converges uniformly on compact subsets of* H_n, *where* Γ *is any discrete subgroup of* $Sp(n, \mathbb{R})$.

Proof

The first statement is just the cocycle relation for Jacobians. In §1 we evaluated the condition $\tilde{m} \in \Phi_n$ to be

$$^t\bar{\tilde{a}}\tilde{a} - {}^t\bar{\tilde{b}}\tilde{b} = \mathbf{1}, \qquad {}^t\bar{\tilde{a}}\tilde{b} \text{ symmetric.} \tag{9}$$

Hence $\tilde{m}$ is unitary if and only if $\tilde{b} = 0$. Now from $|j_m(z)| = 1$ for all $z \in H_n$ we infer $j_m(z)$ to be constant, hence $\tilde{b} = 0$. On the other hand $\tilde{b} = 0$ implies that $\tilde{a}$ is unitary, hence $|j_m(z)| = 1$ identically in z. Finally we deduce from $\tilde{m} = lml^{-1}$ that $\tilde{m}$ is unitary if and only if m is orthogonal. The m appearing in (ii) are called volume-preserving for obvious reasons. The last assertion (iii) is a special case of the following famous theorem due to H. Poincaré [57].

Theorem 3 (H. Poincaré)

Let Γ be a discontinuous group of biholomorphic automorphisms of a bounded domain G in $\mathbb{C}^n$ and denote by $j_\gamma(z)$ the Jacobian of $\gamma \in \Gamma$. Then the Poincaré *series*

$$\sum_{\gamma \in \Gamma} |j_\gamma(z)|^2$$

converges uniformly on any compact subset of G.

Proof

Note that each discontinuous group is enumerable, so the infinite series is well defined. Introduce the norm

$$|z| = \max_{1 \le \nu \le n} |z_\nu|$$

on $\mathbb{C}^n$, where the z_ν denote the components of z.

(i) We show that for any $\varepsilon > 0$ and any compact subset K of G there exists a positive δ such that

$$|\gamma(z) - \gamma(a)| < \varepsilon$$

for all $z \in G$, $a \in K$, $\gamma \in \Gamma$, $|z - a| < \delta$. This statement says that the images with respect to Γ of any polydisc

$$P_\delta(a) = \{z \in \mathbb{C}^n \,|\, |z - a| < \delta\}$$

with center a in K and radius δ remain small uniformly in $\gamma \in \Gamma$ and $a \in K$. It is sufficient to consider a component-function f of γ. Clearly f is bounded by some constant M independent of γ, since G is a bounded domain. Determine r such that the closed polydisc $\overline{P_r(a)}$ belongs completely to G for any $a \in K$. Then by Cauchy's integral formula we have

$$f(z) = \frac{1}{(2\pi i)^n} \int\limits_{\substack{|\zeta_\nu - a_\nu| = r \\ \nu = 1, \ldots, n}} f(\zeta)(\zeta_1 - z_1)^{-1} \ldots (\zeta_n - z_n)^{-1} \, d\zeta_1 \ldots d\zeta_n$$

for any $z \in P_r(a)$. From there we get the estimate

$$|f(z) - f(a)|$$

$$\le \frac{1}{(2\pi)^n} \int\limits_{\substack{|\zeta_\nu - a_\nu| = r \\ \nu = 1, \ldots, n}} M \left| \frac{1}{(\zeta_1 - z_1) \ldots (\zeta_n - z_n)} - \frac{1}{(\zeta_1 - a_1) \ldots (\zeta_n - a_n)} \right| |d\zeta_1| \ldots |d\zeta_n|$$

for arbitrary $a \in K$ and $z \in P_r(a)$. Choose the number $\delta > 0$ independently of $a \in K$ such that the right-hand side becomes less than ε provided $|z - a| < \delta$.

(ii) For each compact subset K of G there exists a number t such that

$$\#\{\gamma \in \Gamma \,|\, z \in \gamma(K)\} \le t$$

no matter what $z \in G$ may be. To prove this assertion let $2r$ be the distance from K to the complement of G, and K_1 the compact set of all points of G with distance at most r from K. Determine $\delta > 0$ by (i) such that

$$|\gamma(z) - \gamma(a)| < r$$

for all $z \in G$, $a \in K$, $\gamma \in \Gamma$, $|z - a| < \delta$. If for some $a \in K$ the image $\gamma(P_\delta(a))$ meets K then $\gamma(a) \in K_1$, hence γ belongs to a finite set (depending on $a \in K$) by discontinuity. But K may be covered by finitely many $P_\delta(a)$, hence

$$\#\{\gamma \in \Gamma \,|\, \gamma(K) \cap K \neq \varnothing\}$$

is finite. This proves the assertion for points $z \in K$, and it follows for the other points of G by the group property.
(iii) Now let K be any compact subset of G and ε any given positive number. Determine K_1 as in (ii). Then (ii) applied to K_1 proves the existence of a number $t > 0$ such that

$$\#\{\gamma \in \Gamma \,|\, z \in \gamma(K_1)\} \leq t$$

for any $z \in G$. Determine the compact subset K_2 such that $K_1 \subset K_2 \subset G$ and the Euclidean volume $V(G - K_2) < \varepsilon/t$. Then $\gamma(K_1) \cap K_2 = \varnothing$ for all but finitely many values of $\gamma \in \Gamma$. If the prime indicates the omission of these finitely many exceptional values we obtain

$$\varepsilon > tV(G - K_2) \geq \sum_{\gamma \in \Gamma}{}' \int_{\gamma(K_1)} \mathrm{d}x\,\mathrm{d}y \geq \sum_{\gamma \in \Gamma}{}' \int_{\gamma(P_r(a))} \mathrm{d}x\,\mathrm{d}y$$

$$= \sum_{\gamma \in \Gamma}{}' \int_{P_r(a)} |j_\gamma(z)|^2 \,\mathrm{d}x\,\mathrm{d}y \geq \sum_{\gamma \in \Gamma}{}' |j_\gamma(a)|^2 (\pi r^2)^n$$

for arbitrary $a \in K$, proving the uniform convergence of the series in question on the compact set K. In the last inequality we used a well-known estimate for holomorphic functions in one complex variable. It may be applied to our situation because the integrals over polydiscs may be understood as iterated one-dimensional integrals.

Corollary

Each discontinuous group of biholomorphic automorphisms of a bounded domain contains at most finitely many volume-preserving mappings.

Now we introduce certain subsets of H_n which are essentially fundamental domains for the underlying discontinuous groups. They were used for instance by I.I. Pjateckij-Šapiro [54] and A. Borel [10] for the modular group.

Definition 3

For any subgroup Γ of $Sp(n, \mathbb{R})$ which acts discontinuously on H_n let

$$B = \{z \in H_n \mid |j_m(z)| \leq 1 \text{ for all } m \in \Gamma\},$$
$$\mathring{B} = \{z \in H_n \mid |j_m(z)| < 1 \text{ for all } m \in \Gamma, m \text{ non-orthogonal}\}.$$

Then the following theorem holds.

Theorem 4

B is closed in H_n, $\mathring{B}$ is the interior of B in H_n, the boundary of B in H_n consists of surfaces $|j_m(z)| = 1$ for non-orthogonal $m \in \Gamma$, and each compact subset of H_n meets only finitely many of these surfaces. B contains at least one representative of each orbit; two points $z_1, z_2 \in B$, at least one of which is contained in $\mathring{B}$ and which belong to the same orbit, are equivalent only with respect to an orthogonal $m \in \Gamma$. Each compact subset of H_n is covered by at most finitely many images of B.

Proof

The geometrical properties of B and $\mathring{B}$ are immediate consequences of statements (ii) and (iii) of Proposition 4. Let z be any point in H_n; then again by Proposition 4 there exists an $m^* \in \Gamma$ such that

$$|j_m(z)| \leq |j_{m^*}(z)| \qquad (m \in \Gamma).$$

Consequently $z^* = m^*\langle z \rangle$ satisfies

$$|j_m(z^*)| = \frac{|j_{mm^*}(z)|}{|j_{m^*}(z)|} \leq 1 \qquad (m \in \Gamma),$$

hence $z^* \in B$. Let $z \in \mathring{B}$, $z^* \in B$ and $z^* = m\langle z \rangle$. Then by the cocycle relation

$$j_{m^{-1}}(z^*) j_m(z) = j_1(z) = 1,$$

hence $|j_m(z)| = 1$ and m turns out to be orthogonal. Finally let K be any compact subset of H_n. If $m\langle K \rangle$ intersects B, there exists a $z \in K$ such that $m\langle z \rangle = z^* \in B$ and we obtain

$$|j_m(z)| = |j_{m^{-1}}(z^*)|^{-1} \geq 1.$$

Since the Poincaré series converges uniformly on K, m belongs to a finite set.

Besides the properties just mentioned in the theorem, almost nothing is known about this fundamental domain. Even for the modular group it is not known whether B is bounded by finitely many algebraic surfaces nor whether B is a fundamental set in the sense of Theorem 1. The important fact that Siegel's fundamental domain is contained in a vertical strip of positive height gets lost with B. One can show for $n = 1$ that B consists of the union of the modular triangle and its image under $z \mapsto -z^{-1}$. For general n the connection between Siegel's fundamental domain and B is not known.

Against this background it is perhaps worthwhile mentioning another geometrical property of B which holds for arbitrary discontinuous groups. Consider the image $\tilde{B} = l\langle B \rangle$ of B with respect to Cayley's transformation.

Proposition 5

For any subgroup Γ of Φ_n operating discontinuously on D_n the fundamental domain $\tilde{B}$ is star-shaped with respect to the origin.

Proof

Star-shaped means that the origin is an interior point of $\tilde{B}$ (which follows from Theorem 4 and (9)) and that for any $w \in \tilde{B}$ the line segment between 0 and w belongs completely to $\tilde{B}$. Now $\tilde{B}$ is the intersection of the subsets

$$E(m) = \{w \in D_n \mid |\det(\bar{b}w + \bar{a})| \geq 1\}, \qquad m = \begin{pmatrix} a & b \\ \bar{b} & \bar{a} \end{pmatrix} \in \Gamma, \qquad b \neq 0.$$

Therefore it is sufficient to show that $E(m)$ is star-shaped for any $m \in \Phi_n$, $b \neq 0$. For this purpose consider the function

$$f(\lambda) = |\det(\lambda\bar{b}w + \bar{a})|^2$$

of the real variable λ for any $w \in D_n$, $m \in \Phi_n$, $b \neq 0$. We will show below that

$$f'(\lambda) < 0 \tag{10}$$

if $0 < \lambda \leq 1$ and $f(\lambda) \leq 1$. Then $E(m)$ turns out to be star-shaped. Indeed, consider any ray originating from 0 and meeting the boundary of $E(m)$ at some point w. Then $f(1) = 1$, and we infer from (10) that λw is an interior point of $E(m)$ for $\lambda < 1$ and fails to belong to $E(m)$ for $\lambda > 1$, λ sufficiently close to 1. Hence there exists at most one boundary point on each ray and $E(m)$ is seen to be star-shaped. In order to prove (10) use the elementary formula

$$\frac{\mathrm{d}}{\mathrm{d}\lambda}\det c = \sigma\left(c^{-1}\frac{\mathrm{d}}{\mathrm{d}\lambda}c\right)\det c$$

for any non-singular matrix c whose entries are differentiable functions of the real variable λ. Abbreviate $q = \lambda\bar{b}w + \bar{a}$ and use $\bar{a}\,{}^t a - \bar{b}\,{}^t b = 1$ in the following computation:

$$f(\lambda)^{-1}f'(\lambda) = \sigma(q^{-1}(2\lambda\bar{b}w\bar{w}\,{}^t b + \bar{a}\bar{w}\,{}^t b + \bar{b}w\,{}^t a)\,{}^t\bar{q}^{-1})$$

$$= \sigma\left(q^{-1}\left(\lambda\bar{b}w\bar{w}\,{}^t b + \frac{q\,{}^t\bar{q}}{\lambda} - \frac{\bar{a}\,{}^t a}{\lambda}\right){}^t\bar{q}^{-1}\right)$$

$$= \frac{1}{\lambda}\sigma(1) - \sigma\left(q^{-1}\bar{b}\left(\frac{1}{\lambda} - \lambda w\bar{w}\right){}^t b\,{}^t\bar{q}^{-1}\right) - \frac{1}{\lambda}\sigma(q^{-1}\,{}^t\bar{q}^{-1}).$$

The second term is negative (note that $b \neq 0$!), hence

$$f(\lambda)^{-1}f'(\lambda) < \frac{1}{\lambda}(n - \sigma(q^{-1}\,{}^t\bar{q}^{-1})) \leq \frac{n}{\lambda}(1 - |\det q|^{-2/n})$$

and the right-hand side is non-positive since $f(\lambda) \leq 1$. The proposition was first proved by E. Gottschling in [29]. Our argument is due to J. Spilker (unpublished).

This section ends with a few remarks on generators and defining relations for the modular group. We mentioned a general method for finding a presentation of Γ_n from the covering of H_n by a fundamental domain and its images under the group. But to determine the presentation explicitly in this way can become difficult. Therefore we prefer an arithmetical procedure. Consider the n embeddings of the elliptic modular group Γ_1 into Γ_n,

$$\Gamma_1 \to \Gamma_n, \qquad \alpha \mapsto m_\alpha^{(\nu)} \qquad (\nu = 1, \ldots, n),$$

where $m_\alpha^{(\nu)}$ is composed of the elements of α at the places (ν, ν), $(\nu, \nu + n)$, $(\nu + n, \nu)$ and $(\nu + n, \nu + n)$ and of the elements of the identity matrix elsewhere. Denote by $M^{(\nu)}$ $(\nu = 1, \ldots, n)$ the images of these homomorphisms and by U the subgroup

$$U = \left\{\begin{pmatrix} u & 0 \\ 0 & {}^t u^{-1} \end{pmatrix} \Big| \, u \text{ unimodular}\right\}$$

of Γ_n. Then we may state the following proposition.

Proposition 6

Γ_n is generated by its subgroups U and $M^{(n)}$, or alternatively by the elements

$$j = \begin{pmatrix} 0 & 1 \\ -1 & 0 \end{pmatrix} \quad \textit{and} \quad \begin{pmatrix} 1 & 0 \\ s & 1 \end{pmatrix},$$

where s runs over any set generating the additive group of symmetric n-rowed integral matrices.

Proof

Note that rows or columns of a modular matrix consist of coprime numbers. Let m be any element of Γ_n and observe the $(n+1)$th column of m while multiplying m from the left. Using elements of $M^{(\nu)}$ as factors we may obtain $m_{\nu,n+1} = 0$ for $\nu = 1, \ldots, n$. Furthermore, taking an appropriate element of U as a factor, the $(n+1)$th column of m becomes the $(n+1)$th unit vector e_{n+1}. Then symplecticity implies that the first row is e_1. By an induction argument we see that Γ_n is generated by

$$U, M^{(1)}, \ldots, M^{(n)} \quad \text{and} \quad \left\{ \begin{pmatrix} 1 & 0 \\ s & 1 \end{pmatrix} \middle| \, s \text{ integral symmetric} \right\}.$$

Obviously $M^{(1)}, \ldots, M^{(n-1)}$, and the elements $\begin{pmatrix} 1 & 0 \\ s & 1 \end{pmatrix}$ are superfluous in this list of generators; thus the first part of the proposition is proven. In the second statement it is surprising that no elements of U appear in the set of generators. Now U is generated by elements which are built up in a simple way by a 2×2 unimodular matrix and the identity matrix. From the generators

$$\begin{pmatrix} 1 & 0 \\ 0 & -1 \end{pmatrix}, \quad \begin{pmatrix} 0 & 1 \\ 1 & 0 \end{pmatrix}, \quad \begin{pmatrix} 1 & 1 \\ 0 & 1 \end{pmatrix}$$

of the unimodular group of degree two and the relation

$$\begin{pmatrix} 5 & 3 \\ 3 & 2 \end{pmatrix} \begin{pmatrix} 0 & 1 \\ 1 & 0 \end{pmatrix} \begin{pmatrix} -3 & 2 \\ 2 & -1 \end{pmatrix} = \begin{pmatrix} 1 & 1 \\ 0 & 1 \end{pmatrix}$$

we infer that U can be generated by symmetric matrices. But for any symmetric unimodular u we have

$$\begin{pmatrix} u^{-1} & 0 \\ 0 & u \end{pmatrix} = \begin{pmatrix} 1 & 0 \\ u & 1 \end{pmatrix} \begin{pmatrix} 0 & 1 \\ -1 & 0 \end{pmatrix} \begin{pmatrix} 1 & 0 \\ u^{-1} & 1 \end{pmatrix} \begin{pmatrix} 0 & 1 \\ -1 & 0 \end{pmatrix} \begin{pmatrix} 1 & 0 \\ u & 1 \end{pmatrix} \begin{pmatrix} 0 & -1 \\ 1 & 0 \end{pmatrix}.$$

Therefore U belongs to the group generated by the second set mentioned in the proposition. A similar situation for $M^{(n)}$ is trivial.

Generators for the modular group were determined by E. Witt [72], L.K. Hua and I. Reiner [32] or in [40]. The second part of the proposition is due to H. Maass [51]. It is much more difficult to determine defining relations. Concerning this question the reader should consult the work of J.E. Humphreys [33], H. Behr [7] and [41], [45]. In this connection we mention the papers of J. Hurrelbrink and U. Rehmann on finite presentations of arithmetically defined groups. F. Kirchheimer and J. Wolfart [38], [39] have applied the method of M. Gerstenhaber and H. Behr to certain Hilbert modular groups.

II

Basic facts on modular forms

4 The linear space of modular forms

For any symplectic map $z \mapsto m\langle z\rangle$ we denote by

$$j(m, z) := \det(cz + d)$$

the $(n + 1)$th root of the reciprocal value of the Jacobian. Of course we have the cocycle relation

$$j(m_1 m_2, z) = j(m_1, m_2\langle z\rangle) j(m_2, z)$$

for arbitrary m_1, $m_2 \in Sp(n, \mathbb{R})$. Modular forms are introduced by the following

Definition

Let Γ_n be Siegel's modular group of degree n, and k any integer. A modular form of degree n and weight k (or synonymously of dimension $-k$) is a complex-valued function f defined on Siegel's half-space H_n which satisfies

(i) *f is holomorphic,*
(ii) *$f(m\langle z\rangle) = j(m, z)^k f(z)$ for all $m \in \Gamma_n$,*
(iii) *f is bounded on Siegel's fundamental domain* ($n = 1$).

First a few remarks concerning this definition. Modular forms of fixed weight and degree obviously form a vector space over $\mathbb{C}$; this is the linear space mentioned in the section title. Condition (iii) is not superfluous as can be seen from the classical absolute modular invariant $j(z)$, which has a simple pole at infinity and is holomorphic elsewhere. The corresponding property for $n > 1$ is a consequence of conditions (i) and (ii), as will soon be seen. One will certainly miss a more general transformation law like

$$f(m\langle z\rangle) = v(m) j(m, z)^k f(z),$$

where k is an arbitrary complex number and $\{v(m) | m \in \Gamma_n\}$ is a family of complex multipliers. Of course $j(m, z)^k$ has to be defined as a certain branch on the simply connected domain H_n. It is well known that for $n = 1$ there exist non-trivial forms of non-integral weights and with non-trivial multipliers. But, for $n > 1$, U. Christian [15] proved that only integral weights may occur. Then multiplier-systems for non-trivial functions become characters of the modular group. By group-theoretical methods it was shown in [50] and [45] that for $n = 2$ only one non-trivial character exists, and none for $n > 2$. Hence our restriction to the trivial multiplier-system and to integral weights excludes only one non-trivial case for $n = 2$ as long as the full modular group of degree $n > 1$ is concerned. The situation becomes different if congruence subgroups are taken into consideration. Nevertheless the only non-trivial multiplier-system for $n = 2$ will sporadically appear in §9.

We first consider those properties of modular forms which hold for the larger class of automorphic forms with respect to the group Δ_n of integral modular substitutions introduced in (3.3). This means we weaken condition (ii) of the definition above to

$$f(z[u] + s) = \det u^k f(z) \tag{1}$$

for all unimodular matrices u and all symmetric integral matrices s. Then $f(z)$ is periodic of period one in each variable z_{kl}. Introducing

$$e^{2\pi i z_{kl}} \qquad (k \leq l)$$

as new variables, f becomes a holomorphic function on a Reinhardt-domain. These are domains which are invariant with respect to multiplying the coordinates by arbitrary complex numbers of modulus one. Such a domain generalizes an annulus in the one-dimensional case. Consequently by Cauchy's integral formula there exists a Laurent expansion of f, which appears as the Fourier series

$$f(z) = \sum_t a(t) e^{2\pi i \sigma(tz)}$$

in the original variables. Here t runs over all half-integral symmetric n-rowed matrices. Half-integral means that the diagonal elements t_k and $2t_{kl}$ $(k \neq l)$ are integers. Note that $\sigma(tz)$ is the most general linear composite of the variables z_{kl} $(k \leq l)$ over $\mathbb{Z}$. This Fourier series is absolutely and uniformly convergent on any compact subset of H_n. By a well-known formula we get the Fourier coefficients back by

$$a(t) = \int_{x \bmod 1} f(z) e^{-2\pi i \sigma(tz)} \, dx,$$

where $dx = \prod_{k \leq l} dx_{kl}$ denotes the Euclidean volume element in the x-space.

By (1) we have for any unimodular u

$$f(z[u]) = \det u^k f(z).$$

Hence we obtain for the Fourier coefficients of f

$$\begin{aligned} a(t[u]) &= \int_{x \bmod 1} f(z)e^{-2\pi i\sigma(t[u]z)}\,dx \\ &= \det u^k \int_{x \bmod 1} f(z[{}^t u])e^{-2\pi i\sigma(tz[{}^t u])}\,dx \\ &= \det u^k a(t). \end{aligned} \tag{2}$$

If kn is odd we may take $u = -\mathbf{1}$ to infer $a(t) = 0$ for all half-integral t. Thus any automorphic form with respect to Δ_n (in particular any modular form) of weight k vanishes identically if $kn \equiv 1 \bmod 2$.

Next we shall prove the boundedness of any modular form in Siegel's fundamental domain if the degree $n > 1$. This phenomenon, which has no analogue in the one-variable case, was discovered by M. Koecher in [47]. In fact it holds for automorphic forms with respect to Δ_n.

Theorem 1 (M. Koecher)
Let f be any automorphic form with respect to Δ_n and $n > 1$. Then f is bounded on any subset

$$\{z \in H_n | y \geq c\mathbf{1}\} \qquad (c > 0)$$

of H_n.

The boundedness in Siegel's fundamental domain follows from Lemma 3.2.

Proof
Consider the Fourier expansion

$$f(z) = \sum_t a(t)e^{2\pi i\sigma(tz)},$$

which converges absolutely on H_n. From the convergence for $z = i\mathbf{1}$ we infer the existence of an $\alpha > 0$ such that

$$|a(t)| \leq \alpha e^{2\pi\sigma(t)} \tag{3}$$

for all half-integral t. We want to show that

$$a(t) = 0 \qquad \text{for } t \ngeq 0.$$

For this purpose take any t which is not positive semi-definite. Then there exists a primitive column v_1 such that

$$t[v_1] < 0.$$

The column v_1 can be completed to a unimodular matrix v. Replacing t by

$t[v]$ and because of (2) we may assume the first diagonal element t_1 of t to be negative. Consider the n-rowed unimodular matrix

$$u = \begin{pmatrix} u^* & 0 \\ 0 & 1 \end{pmatrix}, \qquad u^* = \begin{pmatrix} 1 & m \\ 0 & 1 \end{pmatrix},$$

where u^* is two-rowed and $m \in \mathbb{Z}$. Then by (2) and (3) we have

$$|a(t)| = |a(t[u])| \leq \alpha e^{2\pi\sigma(t[u])}$$

and

$$\sigma(t[u]) = \sigma(t) + t_1 m^2 + 2t_{12}m.$$

But this quantity tends to $-\infty$ for $m \to \infty$, since $t_1 < 0$; hence $a(t) = 0$. Now we may restrict the summation in the Fourier expansion formally to positive semi-definite t

$$f(z) = \sum_{t \geq 0} a(t) e^{2\pi i \sigma(tz)},$$

and this series can be majorized by

$$\sum_{t \geq 0} |a(t)| e^{-2\pi c \sigma(t)} \tag{4}$$

uniformly in z for $y \geq c\mathbf{1}$. This majorant converges and yields a bound for f, since it is the original Fourier series at the point $ic\mathbf{1}$, after having taken absolute values of its terms.

The crucial condition $n > 1$ comes in with the decomposition of u. As can be seen from the proof, the Fourier series of any modular form converges normally on $y \geq c\mathbf{1} (c > 0)$.

If one does not want to use absolute but only plain convergence of the Fourier series, one may argue in the same way but has to prove the convergence of (4) as follows. The convergence of the Fourier series at the point $i(c/2)\mathbf{1}$ implies the estimate

$$|a(t)| \leq \beta e^{\pi c \sigma(t)}$$

for all half-integral t. Then (4) is majorized by

$$\sum_{t \geq 0} e^{-\pi c \sigma(t)}.$$

The number of half-integral positive semi-definite t with $\sigma(t) = d$ is $\mathbf{O}(d^{n(n+1)/2})$, hence

$$\sum_{t \geq 0} e^{-\pi c \sigma(t)} \prec 1 + \sum_{d=1}^{\infty} d^{n(n+1)/2} e^{-\pi c d} < \infty.$$

Koecher's principle was first observed by F. Götzky [27] in the case of Hilbert modular forms related to the number field $\mathbb{Q}(\sqrt{5})$; but the importance of his result was obviously not recognized at that time. Another essential interpretation of this principle can be given within the framework of Satake's compactification.

From now on we apply the transformation law (ii) (cf. the definition at the beginning of this section) for all of Γ_n.

Corollary

Each modular form of negative weight vanishes identically.

Proof

Let f be any modular form of weight $k < 0$. Then f is bounded in Siegel's fundamental domain F_n by Theorem 1 for $n > 1$, respectively by definition for $n = 1$. The quantity $\det y^{k/2}$ is bounded in F_n by Lemma 3.2, since k is negative. Hence, because of its invariance with respect to Γ_n,

$$\det y^{k/2}|f(z)| \leq K$$

everywhere in H_n with a certain constant K. Now the Fourier coefficients of $f(z)$ satisfy

$$a(t)\mathrm{e}^{-2\pi\sigma(ty)} = \int_{x \bmod 1} f(z)\mathrm{e}^{-2\pi i\sigma(tx)}\,\mathrm{d}x,$$

$$|a(t)|\mathrm{e}^{-2\pi\sigma(ty)} \leq \sup_{x \bmod 1} |f(x+iy)| \leq K \det y^{-k/2}.$$

Let y tend to zero, to obtain $a(t) = 0$ for any $t \geq 0$.

Consider the vector space M_n^k of all modular forms of fixed weight k and degree n. Our next goal will be concerned with the dimension of this linear space. Beyond finiteness we at least want a bound for the dimension which depends on k in such a way that algebraic relations in the graded ring of all modular forms can be deduced. There are different proofs, some of which use the maximum principle in one form or the other. For instance, H. Maass [49] considered the Fourier expansion of a modular form. He proved that a modular form is already zero if sufficiently many Fourier coefficients vanish. M. Eichler improved his procedure in [18] working with Fourier–Jacobi expansions, which will be considered in §8 in another context. We present two methods in this book. The first one is due to I.I. Pjateckij-Šapiro [54] and covers a more general situation; another one in §6 depends on the theory of integral equations.

Proposition 1

To any real number $c > 0$ there exists a compact set K contained in Siegel's fundamental domain F_n and a positive constant α such that

$$\sup_{y \geq c1} |f| \leq \alpha^k \sup_K |f|$$

for all modular forms f of weight k. The number α is independent of f and k, but may depend on c and n.

Proof

By Lemma 3.2 we may assume from the beginning c to be so small that

$$F_n \subset \{z \in H_n | y \geq c\mathbf{1}\}. \tag{5}$$

We sharpen the argument in the proof of Koecher's theorem. Let

$$f(z) = \sum_{t \geq 0} a(t) e^{2\pi i \sigma(tz)}$$

be the Fourier series of any modular form of weight k. Then estimate the coefficients by

$$a(t) = \int_{x \bmod 1} f\left(x + i\frac{c}{2}\mathbf{1}\right) e^{-2\pi i \sigma(t(x + i(c/2)\mathbf{1}))} \, dx,$$

$$|a(t)| \leq \sup_{x \bmod 1} \left| f\left(x + i\frac{c}{2}\mathbf{1}\right)\right| e^{\pi c \sigma(t)}.$$

Hence we obtain from the Fourier expansion

$$|f(z)| \leq \alpha_1 \sup_{x \bmod 1} \left| f\left(x + i\frac{c}{2}\mathbf{1}\right)\right|, \tag{6}$$

for any $z \in H_n$, $y \geq c\mathbf{1}$, where the constant

$$\alpha_1 = \sum_{t \geq 0} e^{-c\pi\sigma(t)}$$

depends only on n and c. By Theorem 3.2 we may determine a compact subset K of F_n and finitely many m_ν $(\nu = 1, \ldots, l)$ in Γ_n such that

$$\left\{x + i\frac{c}{2}\mathbf{1} | x \bmod 1\right\} \subset \bigcup_{\nu=1}^{l} m_\nu\langle K\rangle.$$

Hence by the transformation law of modular forms we have

$$\sup_{x \bmod 1} \left| f\left(x + i\frac{c}{2}\mathbf{1}\right)\right| \leq \alpha_2^k \sup_K |f|, \tag{7}$$

where

$$\alpha_2 = \sup_{1 \leq \nu \leq l, z \in K} |\det(c_\nu z + d_\nu)|$$

is again independent of f and k. We combine (6) and (7) to obtain

$$\sup_{y \geq c\mathbf{1}} |f| \leq \alpha_1 \alpha_2^k \sup_K |f|.$$

Note that $\alpha_1 \geq 1$, therefore the assertion becomes true with $\alpha = \alpha_1\alpha_2$, if $k > 0$. If $k = 0$ we only have available

$$\sup_{y \geq c\mathbf{1}} |f| \leq \alpha_1 \sup_K |f|$$

for any modular form f of weight zero. Replace f by f^t ($t = 1, 2, \ldots$) in this inequality and take tth roots on both sides. Then

$$\sup_{y \geq c\mathbf{1}} |f| \leq \alpha_1^{1/t} \sup_K |f|.$$

Finally let t tend to infinity to obtain the assertion in the remaining case $k = 0$ as well.

As an application choose c again as in (5); then for any modular form f of weight zero we obtain

$$\sup_{y \geq c\mathbf{1}} |f| \leq \sup_K |f| \leq \sup_{y \geq c\mathbf{1}} |f|.$$

Hence equality holds and f turns out to be constant by the maximum principle.

Corollary

Any modular form of weight zero is constant.

In the next proposition we will estimate the dimension of linear spaces of holomorphic functions in a rather general situation. We consider arbitrary domains in the complex number space $\mathbb{C}^n$ equipped with the norm

$$|z| = \max_{1 \leq \nu \leq n} |z_\nu|$$

and use the same notation as in the proof of Poincaré's theorem in §3.

Proposition 2

Let A be any domain in $\mathbb{C}^n$, V a vector space of holomorphic functions on A, and B any compact subset of A with non-empty interior. Assume

$$\sup_A |f| \leq \beta \sup_B |f| \tag{8}$$

for all $f \in V$ with some constant $\beta \geq 2$ independent of f. Then there exists a number γ depending only on A and B such that

$$\dim V \leq \gamma (\log \beta)^n.$$

Proof

Choose $\rho > 0$ such that any polydisc $P_{2\rho}(z)$ with radius 2ρ and center z in B is contained in A. Then cover the compact set B by finitely many polydiscs of radius ρ:

$$B \subset \bigcup_{\nu=1}^{q} P_\rho(z_\nu), \qquad z_1, \ldots, z_q \in B. \tag{9}$$

Obviously ρ and q only depend on A and B. First we will show that any

$f \in V$ with zeros of order $> \log\beta/\log 2$ in $z_1, \ldots, z_q$ vanishes identically. Indeed, let f be any such function, put

$$m = \left[\frac{\log\beta}{\log 2} + 1\right],$$

and choose a point $z_0 \in B$ for which

$$\sup_B |f| = |f(z_0)|.$$

Then $z_0 \in P_\rho(z_\mu)$ for at least one μ $(1 \leq \mu \leq q)$ by (9) and

$$z_\mu + \lambda(z_0 - z_\mu) \in A$$

for $|\lambda| \leq 2$. Consider the function

$$\varphi(\lambda) = f(z_\mu + \lambda(z_0 - z_\mu))\lambda^{-m}$$

of the single complex variable λ. It is holomorphic on $|\lambda| \leq 2$, since f vanishes of order $\geq m$ in z_μ. Therefore by the maximum principle there exists a λ^* with $|\lambda^*| = 2$ such that

$$|\varphi(\lambda^*)| = \sup_{|\lambda| \leq 2} |\varphi|.$$

Put $z^* = z_\mu + \lambda^*(z_0 - z_\mu)$, then $z^* \in A$ and

$$\sup_B |f| = |f(z_0)| = |\varphi(1)| \leq |\varphi(\lambda^*)| = \left|\frac{f(z^*)}{2^m}\right|.$$

We combine this inequality with the assumption (8) to infer

$$2^m \sup_B |f| \leq |f(z^*)| \leq \sup_A |f| \leq \beta \sup_B |f|.$$

But $2^m > \beta$ by the choice of m, hence $f = 0$. The number of partial derivatives of a function in n variables up to order m is $\binom{n+m}{m}$. If there were more than

$$l = q\binom{n+m-1}{m-1}$$

linearly independent functions in V, then by solving a system of l linear equations in more than l unknowns we would be able to find a non-trivial linear composite of those functions with zeros of order $\geq m$ in $z_1, \ldots, z_q$, contradicting the knowledge we have just gained before. Hence

$$\dim V \leq q\binom{n+m-1}{m-1} \leq qm^n \leq q\left(\frac{\log\beta}{\log 2} + 1\right)^n \leq \gamma(\log\beta)^n,$$

with a constant γ depending only on A and B.

Enclose Siegel's fundamental domain in a vertical strip of positive height and apply Propositions 1 and 2. Then we immediately obtain an upper bound for the dimension of M_n^k for $k > 0$. Summarizing with former results for $k \leq 0$ we may state

Theorem 2

The dimension of the linear space M_n^k of all modular forms of weight k and degree n satisfies

$$\dim M_n^k \begin{cases} < d_n k^{n(n+1)/2} & \text{for } k > 0, \\ = 1 & \text{for } k = 0, \\ = 0 & \text{for } k < 0, \end{cases}$$

where the constant d_n only depends on n.

The existence of modular forms which are different from zero is by no means trivial, and we refer to subsequent parts of this book for this question. At present we just mention a few results in advance without claiming completeness. We know already that there are no forms different from zero if nk is odd or $k < 0$. If k is even and $k > n + 1$, non-trivial forms do exist, for instance those given by Eisenstein series. If k is odd and sufficiently large and n is even, one may construct non-trivial forms by Poincaré series. Finally, if k is positive and small, non-trivial forms only exist if $k \equiv 0 \bmod 4$; they are represented by theta-series.

The most important application of Theorem 2 is concerned with the algebraic dependence of sufficiently many modular forms of arbitrary weights. If $f_1, \ldots, f_h$ are modular forms of weights $k_1, \ldots, k_h$, respectively, then we may restrict ourselves to such algebraic equations

$$A(f_1, \ldots, f_h) = 0,$$

where A is an isobaric polynomial with respect to the weights $k_1, \ldots, k_h$; i.e. the monomials $f_1^{x_1} \ldots f_h^{x_h}$ which appear in such an equation have to satisfy

$$k_1 x_1 + \cdots + k_h x_h = t$$

for some fixed t. Indeed, if $f_1, \ldots, f_h$ satisfy any algebraic equation, then its isobaric parts already have to vanish. Otherwise one would obtain a non-trivial algebraic equation for $\det(cz + d)$ on replacing z by $m\langle z\rangle$, since different isobaric parts attain different powers of $\det(cz + d)$ as factor. This contradicts the fact that $\det(cz + d)$ attains infinitely many values for fixed z and variable $m \in \Gamma_n$.

In order to find an isobaric algebraic equation for modular forms $f_1,\ldots,f_h$ of positive weights $k_1,\ldots,k_h$, we have to estimate the number of terms in such an equation from below. Put $K = k_1 k_2 \ldots k_h$ and let μ be any positive integer. Then the number of terms in an isobaric polynomial of weight μK can be estimated trivially by

$$\begin{aligned}&\#\{(x_1,\ldots,x_h)\in\mathbb{Z}^h | x_\nu \geq 0; k_1x_1+\cdots+k_hx_h=\mu K\}\\ &\geq \#\{(x_1,\ldots,x_h)\in\mathbb{Z}^h | x_\nu\geq 0; x_1+\cdots+x_h=\mu\} \qquad (10)\\ &\geq \frac{1}{(h-1)!}\mu^{h-1}.\end{aligned}$$

If this number exceeds the dimension of the space of modular forms of weight μK, then the functions $f_1^{x_1}\ldots f_h^{x_h}$ become linearly dependent, i.e. $f_1,\ldots,f_h$ are algebraically dependent. Combining the estimates in (10) and Theorem 2 we obtain the condition

$$d_n(\mu K)^{h-2} < \frac{1}{(h-1)!}\mu^{h-1}, \qquad (11)$$

where $h = n(n+1)/2 + 2$. This inequality is true for sufficiently large μ. Hence we have proved that each family of $n(n+1)/2 + 2$ modular forms of arbitrary weights is algebraically dependent.

The weight of the isobaric equation for $f_1,\ldots,f_h$ just obtained is μK, but the magnitude of μ depends on $k_1,\ldots,k_h$. In chapter V we want to prove that the field of modular functions is an algebraic function field. For this purpose it will turn out to be essential that μ can be chosen independently of $k_1,\ldots,k_h$. Therefore we improve the estimate (10) by the following technically more complicated considerations.

Lemma

Let $h, \mu, k_1,\ldots,k_h$ be positive integers, $2h-2|\mu$, $h\geq 2$ and $K = k_1k_2\ldots k_h$. Then the order of the set

$$X = \{(x_1,\ldots,x_h)\in\mathbb{Z}^h | x_\nu\geq 0; k_1x_1+\cdots+k_hx_h=\mu K\}$$

satisfies

$$\#(X) > \left(\frac{\mu}{2h-2}\right)^{h-1} K^{h-2}.$$

Proof

Consider the set L of $(h-1)$-dimensional integral vectors $\zeta = (\zeta_1,\ldots,\zeta_{h-1})$ such that

$$0\leq \zeta_\nu \leq \frac{\mu K}{(2h-2)k_\nu} \qquad (\nu = 1,\ldots,h-1).$$

Obviously we have

$$\#(L) \geq 1 + \left(\frac{\mu}{2h-2}\right)^{h-1} k_h K^{h-2}. \tag{12}$$

Distribute the elements of L according to the behavior of $k_1\zeta_1 + \cdots + k_{h-1}\zeta_{h-1}$ modulo k_h. Then there exists at least one residue class $\Delta \bmod k_h$ such that

$$S = \{\zeta \in L | k_1\zeta_1 + \cdots + k_{h-1}\zeta_{h-1} \in \Delta\}$$

is of order

$$\#(S) \geq \frac{\#(L)}{k_h}. \tag{13}$$

Fix an element $\xi \in S$, of which we may assume

$$0 \leq \xi_\nu < k_h \qquad (\nu = 1, \ldots, h-1).$$

Then we define a map of S into X by the assignment $\zeta \mapsto x$, where

$$x_\nu = \zeta_\nu + k_h - \xi_\nu \qquad (\nu = 1, \ldots, h-1),$$

$$x_h = \mu\frac{K}{k_h} - \frac{1}{k_h}(k_1 x_1 + \cdots + k_{h-1}x_{h-1}).$$

To check $x \in X$ is obvious up to the verification of $x_h \geq 0$: we have

$$\begin{aligned} x_h &\geq \mu\frac{K}{k_h} - \frac{1}{k_h}(k_1(\zeta_1 + k_h) + \cdots + k_{h-1}(\zeta_{h-1} + k_h)) \\ &\geq \frac{\mu K}{2k_h} - (k_1 + \cdots + k_{h-1}) \\ &\geq \frac{(h-1)K}{k_h} - (k_1 + \cdots + k_{h-1}) \geq 0. \end{aligned}$$

The map $\zeta \mapsto x$ is injective, hence by (12) and (13)

$$\#(X) \geq \#(S) \geq \frac{\#(L)}{k_h} > \left(\frac{\mu}{2h-2}\right)^{h-1} K^{h-2}.$$

In order to get an isobaric algebraic equation for h modular forms we now have to satisfy instead of (11) the condition

$$d_n(\mu K)^{h-2} \leq \left(\frac{\mu}{2h-2}\right)^{h-1} K^{h-2}, \qquad 2h-2 | \mu,$$

which means

$$d_n(n(n+1)+2)^{n(n+1)/2+1} \leq \mu, \qquad n(n+1)+2 | \mu.$$

Hence we can choose a number μ which depends only on n.

Theorem 3

Let $h = n(n+1)/2 + 2$. Any set of h modular forms $f_1, \ldots, f_h$ of positive weights $k_1, \ldots, k_h$ satisfies an isobaric algebraic equation

$$A(f_1, \ldots, f_h) = 0$$

of total weight $\mu k_1 \ldots k_h$, where the integer μ depends only on n.

In recent years there has been much activity to find explicit formulas for the dimension of M_n^k. Selberg's trace formula is one of the key methods for this program. But at present final results are only known for degrees $n \leq 3$. We refer to K.I. Hashimoto [30] for $n = 2$ and S. Tsuyumine [67] for $n = 3$, besides numerous other papers by U. Christian, M. Eie, J.-I. Igusa, Y. Morita and R. Tsushima on this interesting subject.

5 Eisenstein series and the Siegel operator

Already at the very beginning of the theory C.L. Siegel [62] related modular forms in many variables to those in fewer variables. He considered a certain linear map Φ between modular forms of degree n and forms of degree $n-1$, the weight k remaining fixed. We introduce this so-called Siegel operator in a slightly generalized manner.

Proposition 1

Let n, r be integers, $n \geq 1, 0 \leq r \leq n$. Consider sequences $z^{(\nu)}$ $(\nu = 1, 2, \ldots)$ in H_n,

$$z^{(\nu)} = \begin{pmatrix} z^* & z_2^{(\nu)} \\ {}^t z_2^{(\nu)} & z_4^{(\nu)} \end{pmatrix},$$

such that $z^* \in H_r$ is fixed, $z_2^{(\nu)}$ is bounded and all the eigenvalues of $y_4^{(\nu)} = \operatorname{Im} z_4^{(\nu)}$ tend to infinity. Then for any $f \in M_n^k$ the limit

$$f|\Phi^{n-r}(z^*) = \lim_{\nu \to \infty} f(z^{(\nu)})$$

exists and represents a modular form of degree r and weight k. Φ^{n-r} is the $(n-r)$th iteration of Φ.

Remark

We use the convention

$$M_0^k = \begin{cases} \mathbb{C} & \text{for } k \geq 0 \\ 0 & \text{for } k < 0. \end{cases}$$

Proof

From the decomposition (2.3)

$$y = \begin{pmatrix} y^* & y_2 \\ {}^t y_2 & y_4 \end{pmatrix} = \begin{pmatrix} y^* - y_4^{-1}[{}^t y_2] & 0 \\ 0 & y_4 \end{pmatrix}\left[\begin{pmatrix} 1 & 0 \\ y_4^{-1}\,{}^t y_2 & 1 \end{pmatrix}\right]$$

we see that $z^{(\nu)}$ $(\nu = 1, 2, \ldots)$ is located in a subset $y \geq c\mathbf{1}$ $(c > 0)$ of H_n, where the Fourier series of f,

$$f(z) = \sum_{t \geq 0} a(t) e^{2\pi i \sigma(tz)},$$

converges uniformly. Hence we may pass to the limit term by term and obtain

$$f|\Phi^{n-r}(z^*) = \sum_{t^* \geq 0} a\begin{pmatrix} t^* & 0 \\ 0 & 0 \end{pmatrix} e^{2\pi i \sigma(t^* z^*)}, \tag{1}$$

where t^* runs over all r-rowed half-integral positive semi-definite matrices. First we remark that the limit does not depend on the special choice of the sequence $z^{(\nu)}$. Then the limiting process is of a very simple kind, since the terms of the original Fourier series either tend to zero for $\nu \to \infty$ or do not depend on ν at all. Since the series (1) again converges uniformly on compact subsets of H_r, it represents a function $f|\Phi^{n-r}$ holomorphic on H_r (and bounded in the modular triangle for $r = 1$). For $m^* \in \Gamma_r$, $z^* \in H_r$ and $\lambda > 0$ set

$$z = \begin{pmatrix} z^* & 0 \\ 0 & i\lambda\mathbf{1} \end{pmatrix}, \qquad m = \begin{pmatrix} a & b \\ c & d \end{pmatrix}, \tag{2}$$

where

$$a = \begin{pmatrix} a^* & 0 \\ 0 & \mathbf{1} \end{pmatrix}, \quad b = \begin{pmatrix} b^* & 0 \\ 0 & 0 \end{pmatrix}, \quad c = \begin{pmatrix} c^* & 0 \\ 0 & 0 \end{pmatrix}, \quad d = \begin{pmatrix} d^* & 0 \\ 0 & \mathbf{1} \end{pmatrix}.$$

Then $m \in \Gamma_n$, $z \in H_n$ and the transformation law of $f \in M_n^k$ implies

$$f\begin{pmatrix} m^*\langle z^* \rangle & 0 \\ 0 & i\lambda\mathbf{1} \end{pmatrix} = j(m^*, z^*)^k f\begin{pmatrix} z^* & 0 \\ 0 & i\lambda\mathbf{1} \end{pmatrix}.$$

Let λ tend to infinity to infer

$$f|\Phi^{n-r}(m^*\langle z^* \rangle) = j(m^*, z^*)^k f|\Phi^{n-r}(z^*).$$

Hence $f|\Phi^{n-r}$ is a modular form of degree r and weight k. The interpretation of Φ^{n-r} as an iteration is immediate from (1).

Of course the simplest way to describe the above limit is

$$f|\Phi^{n-r}(z^*) = \lim_{\lambda \to \infty} f\begin{pmatrix} z^* & 0 \\ 0 & i\lambda\mathbf{1} \end{pmatrix} \qquad (z^* \in H_r), \tag{3}$$

and usually we shall only work with this formula as a definition. The mapping

$$\Phi\colon M_n^k \to M_{n-1}^k,$$

being linear, suggests the investigation of its kernel and image.

Definition 1

A modular form f is called a cusp form if $f|\Phi = 0$. The set of all cusp forms of fixed weight k and degree n is a subspace S_n^k of M_n^k, which is the kernel of Φ.

For $n = 0$ we add $S_0^k = M_0^k$ as a definition. Cusp forms can be introduced in this simple way only for the full modular group. For congruence subgroups, for instance, further conditions have to be satisfied.

We want to characterize cusp forms by different properties. The first one is concerned with its Fourier series.

Proposition 2

Let $f \in M_n^k$ and $n \geq 1$. Then f is a cusp form if and only if

$$f(z) = \sum_{t>0} a(t)e^{2\pi i\sigma(tz)},$$

where t runs over all half-integral positive matrices.

Proof

From (1) we deduce that f is a cusp form if and only if $a(t) = 0$ for each t of the shape

$$t = \begin{pmatrix} t_1 & 0 \\ 0 & 0 \end{pmatrix}.$$

Now for any singular t there exists a unimodular u such that $t[u]$ is of this form and vice versa; furthermore we have

$$a(t) = \pm a(t[u])$$

by (4.2). Hence f is a cusp form iff $a(t) = 0$ for any singular t.

Another characterization of cusp forms states that f approaches zero exponentially if y tends in a certain way to infinity.

Proposition 3

Let f be a cusp form and let $c > 0$. Then there exist positive numbers c_1, c_2 such that

$$|f(z)| \leq c_1 e^{-c_2\sigma(y)}$$

for all $z \in H_n$, for which y is reduced (or in $Q_n(t)$) and $y \geq c\mathbf{1}$.

Proof
We obtain from the Fourier expansion

$$|f(z)| \leq \sum_{t>0} |a(t)| e^{-2\pi\sigma(ty)}$$

for all z in H_n. Now we estimate the exponential term in two different ways. On the one hand we have

$$\sigma(ty) \geq c\sigma(t)$$

for $y \geq c\mathbf{1}$; on the other hand y is assumed to be reduced in Minkowski's sense. Therefore we may replace y by its diagonalization according to Lemma 2.2. Since the diagonal elements of t are positive integers we get

$$\sigma(ty) \geq c_3\sigma(y)$$

with a positive constant c_3 depending only on n. Hence combining our estimates we obtain

$$\begin{aligned} |f(z)| &\leq e^{-\pi c_3\sigma(y)} \sum_{t>0} |a(t)| e^{-\pi c\sigma(t)} \\ &\leq c_1 e^{-c_2\sigma(y)}. \end{aligned}$$

Remark
It is easily checked that the converse of Proposition 3 also holds. In this context notice that for any $y \in R_{n-1}$ the complemented matrix

$$\begin{pmatrix} y & 0 \\ 0 & \lambda \end{pmatrix} \in R_n,$$

if λ is sufficiently large. This fact can immediately be seen from Minkowski's reduction conditions.

Finally we show a typical behavior for cusp forms in all of H_n.

Proposition 4
Let f be a modular form of weight k and degree n. Then f is a cusp form if and only if there exist positive numbers a_1, a_2 such that

$$\det y^{k/2} |f(z)| \leq a_1 e^{-a_2 \det y^{1/n}} \tag{4}$$

for all $z \in H_n$.

Proof

We may assume $k \geq 0$, since otherwise $f = 0$. First let f be a cusp form. For any given $z \in H_n$ determine $m \in \Gamma_n$ such that $z^* = m\langle z \rangle$ belongs to Siegel's fundamental domain. Then z^* satisfies the assumptions of Proposition 3 and the left-hand side of the inequality (4) is invariant with respect to Γ_n. Hence we obtain

$$\det y^{k/2} |f(z)| \leq c_1 \det y^{*k/2} e^{-c_2 \sigma(y^*)}.$$

Now we have

$$\det y^{*1/n} \leq \frac{\sigma(y^*)}{n}$$

by the inequality of the geometric and arithmetic means, and $\det y \leq \det y^*$ since y^* has maximal height in its orbit. So we can continue the estimate as follows:

$$\begin{aligned} \det y^{k/2} |f(z)| &\leq c_1 \det y^{*k/2} e^{-c_2 n \det y^{*1/n}} \\ &\leq a_1 e^{-a_2 \det y^{*1/n}} \\ &\leq a_1 e^{-a_2 \det y^{1/n}}. \end{aligned}$$

To prove the converse apply the Φ-operator to both sides of the inequality (4) to obtain $f | \Phi = 0$.

From Proposition 4 we deduce that a cusp form f satisfies

$$f(m\langle z \rangle) \det(cz + d)^{-k} | \Phi = 0$$

not only for $m \in \Gamma_n$ but for any $m \in Sp(n, \mathbb{R})$. Indeed, for $k \leq 0$ any cusp form is zero, and for $k > 0$

$$\det y^{k/2} |f(m\langle z \rangle)| |\det(cz + d)|^{-k} = \det(\operatorname{Im} m\langle z \rangle)^{k/2} |f(m\langle z \rangle)|$$

is bounded by (4). Now apply the Φ-operator in its original form as a limit to both sides of this equation to obtain our result. This observation offers a hint of how to introduce cusp forms for other subgroups of $Sp(n, \mathbb{R})$.

Now we turn to the image of Φ, in particular we ask whether Φ is surjective. An affirmative answer to this question was first given by H. Maass [49] for large and even values of k. He used for his proof a certain type of Poincaré series. Here we work with Eisenstein series to boundary components by which the whole theory is simplified considerably. These Eisenstein series were first introduced in [43]; they are also related to the work of R.P. Langlands [48] and W.L. Baily and A. Borel [5].

The surjectivity of

$$\Phi \colon M_n^k \to M_{n-1}^k$$

is guaranteed for some k if

$$S_r^k \subset \Phi^{n-r}(M_n^k) \tag{5}$$

is valid for $0 \leq r \leq n$. In fact, then we can deduce

$$M_r^k \subset \Phi^{n-r}(M_n^k) \tag{6}$$

for $0 \leq r \leq n$ from (5) by induction on r. If $r = 0$, (6) is identical with (5) since $M_0^k = S_0^k$. Let (6) hold for $r - 1$ and take any $f \in M_r^k$. Then $f|\Phi \in M_{r-1}^k$ and by the induction hypothesis there exists an $F \in M_n^k$ such that $f|\Phi = F|\Phi^{n-r+1}$. Hence $f - F|\Phi^{n-r}$ is a cusp form and is contained in $\Phi^{n-r}(M_n^k)$ by (5). So we obtain $f \in \Phi^{n-r}(M_n^k)$.

Having (5) in mind we must lift cusp forms with respect to Φ. This is a non-trivial question, since the number of variables must be enlarged. The formation of Eisenstein series to boundary components will turn out to be an adequate method as long as the weight k is even and sufficiently large. In this context certain subgroups of Γ_n will appear which have a geometric meaning and which shall be considered first.

Let D_n be the unit-circle of degree n introduced in Definition 1.3. Its boundary ∂D_n consists of all points w, for which the eigenvalues of $w\bar{w}$ are located between 0 and 1, and at least one eigenvalue is 1. Following I.I. Pjateckij-Šapiro [56] a connected subset T of ∂D_n is called a boundary component if T is a complex submanifold in $\mathbb{C}^{n(n+1)/2}$ and if any holomorphic curve, which is contained entirely within ∂D_n and intersects T, is completely contained in T. By a holomorphic curve we mean here a complex curve which can be represented by a holomorphic vector-valued function

$$\varphi(t) = (\varphi_1(t), \ldots, \varphi_p(t)) \qquad \left(p = \frac{n(n+1)}{2}\right)$$

of a complex parameter t in a disc $|t| < \delta$. For $n = 1$ the boundary components of the unit-circle are obviously the single points w of the circumference.

For arbitrary n consider the subsets

$$D_{n,r} = \left\{\begin{pmatrix} w & 0 \\ 0 & 1 \end{pmatrix} \,\middle|\, w \in D_r\right\} \qquad (0 \leq r \leq n-1)$$

of ∂D_n. These are complex submanifolds of $\mathbb{C}^{n(n+1)/2}$. In order to prove $D_{n,r}$ to be a boundary component of D_n we have to verify the second condition above concerned with holomorphic curves. For this purpose we need the following simple lemma.

Lemma 1

Let $\varphi_1(t), \dots, \varphi_p(t)$ be holomorphic functions of the single complex variable t in $|t| < \delta$. Then

$$\sum_{v=1}^{p} |\varphi_v(0)|^2 = \sup_{|t|<\delta} \sum_{v=1}^{p} |\varphi_v(t)|^2$$

if and only if all the φ_v are constant.

Proof

By Cauchy's integral formula we have

$$\varphi_v^2(0) = \frac{1}{2\pi} \int_0^{2\pi} \varphi_v^2(\rho e^{i\tau})\, d\tau \qquad (0 < \rho < \delta).$$

Hence

$$\sum_v |\varphi_v(0)|^2 \le \frac{1}{2\pi} \int_0^{2\pi} \sum_v |\varphi_v(\rho e^{i\tau})|^2\, d\tau \le \sup_{|t|<\delta} \sum_v |\varphi_v(t)|^2,$$

and the sign of equality only holds if

$$|\varphi_v(0)|^2 = \frac{1}{2\pi} \int_0^{2\pi} |\varphi_v(\rho e^{i\tau})|^2\, d\tau$$

for all v. From there we infer φ_v to be constant.

Consider now an arbitrary holomorphic curve $w(t)$ satisfying

$$w(t) \in \partial D_n, \qquad w(0) \in D_{n,r} \qquad (|t| < \delta). \tag{7}$$

We have to prove that $w(t)$ is completely contained in $D_{n,r}$. Decompose

$$w(t) = \begin{pmatrix} w_1 & w_2 \\ {}^t w_2 & w_4 \end{pmatrix},$$

where w_1 is r-rowed. Then by (7)

$$1 - {}^t\bar{w}_2 w_2(t) - {}^t\bar{w}_4 w_4(t) \ge 0 \qquad (|t| < \delta),$$

and forming the trace we obtain

$$\sum_{\substack{k=1,\dots,n \\ l=r+1,\dots,n}} |w_{kl}(t)|^2 \le n - r \qquad (|t| < \delta).$$

The supremum of the left-hand side is $n - r$ and is attained for $t = 0$, since $w(0) \in D_{n,r}$. Therefore we can apply the lemma to conclude

$$w_2(t) = 0, \qquad w_4(t) = 1$$

for all t in the disc $|t| < \delta$. Finally we must show that $\bar{w}_1 w_1(t) < 1$ for $|t| < \delta$. From (7) we have available

$$\bar{w}_1 w_1(t) \le 1 \quad (|t| < \delta), \qquad \bar{w}_1 w_1(0) < 1. \tag{8}$$

Therefore it is sufficient to show that

$$\det(\mathbf{1} - \bar{w}_1 w_1(t)) \neq 0$$

for all $|t| < \delta$. Assume to the contrary $\mathbf{1} - \bar{w}_1 w_1(t)$ were singular for some $t = t_0$. Determine the column $\xi_0 \neq 0$ such that

$$(\mathbf{1} - \bar{w}_1 w_1(t_0))\{\xi_0\} = 0 \tag{9}$$

and consider the function

$$\xi(t) = w_1(t)\xi_0 \qquad (|t| < \delta).$$

Then by (8) and (9)

$${}^t\bar{\xi}\xi(t) \leq {}^t\bar{\xi}_0\xi_0 \quad (|t| < \delta), \qquad {}^t\bar{\xi}\xi(0) < {}^t\bar{\xi}_0\xi_0, \qquad {}^t\bar{\xi}\xi(t_0) = {}^t\bar{\xi}_0\xi_0,$$

contradicting the lemma above. Thus we have shown that the subsets $D_{n,r}$ $(0 \leq r \leq n-1)$ are boundary components.

It can be proved that any boundary component of D_n can be mapped onto one of the $D_{n,r}$ by a symplectic transformation. Therefore we call $D_{n,r}$ the standard boundary component of degree r. For convenience we add $D_{n,n} := D_n$ as the (improper) standard boundary component of degree n. We have defined boundary components by analytical properties. It should be mentioned that there exist different ways to introduce this concept, for instance one may characterize boundary components by metrical properties. For this purpose one has to extend the symplectic metric to the boundary of D_n as a pseudometric. For details the reader may consult [56].

We want to determine the fixed groups of the standard boundary components in the symplectic group. This question can be discussed better in the bounded realization D_n of H_n since then the corresponding mappings can be extended canonically to the boundary according to Proposition 1.4. For $n \geq 1, 0 \leq r \leq n$ let

$$\hat{C}_{n,r} = \{m \in \Phi_n \,|\, m\langle D_{n,r}\rangle = D_{n,r}\}$$

be the fixed group of $D_{n,r}$ and

$$\hat{B}_{n,r} = \{m \in \Phi_n \,|\, m\langle w\rangle = w \text{ for all } w \in D_{n,r}\}$$

the group of all $m \in \Phi_n$, which leave $D_{n,r}$ pointwise fixed. Set $m = \begin{pmatrix} a & b \\ \bar{b} & \bar{a} \end{pmatrix}$ and decompose

$$a = \begin{pmatrix} a_1 & a_2 \\ a_3 & a_4 \end{pmatrix}, \qquad b = \begin{pmatrix} b_1 & b_2 \\ b_3 & b_4 \end{pmatrix},$$

where a_1, b_1 are r-rowed. Then, as a straightforward calculation shows, $m \in \hat{C}_{n,r}$ if and only if

$$a_2 + b_2 = 0, \qquad \bar{a}_3 = b_3, \qquad a_4 + b_4 \text{ real},$$

and $m \in \hat{B}_{n,r}$, iff moreover

$$a_1 = \pm 1, \qquad b_1 = 0.$$

We pass from D_n to H_n by Cayley's transformation according to Proposition 1.2. Then we have to replace $\hat{C}_{n,r}$ and $\hat{B}_{n,r}$ by

$$C_{n,r} = l^{-1}\hat{C}_{n,r}l \quad \text{and} \quad B_{n,r} = l^{-1}\hat{B}_{n,r}l.$$

These groups will also be called fixed groups of the boundary component. If

$$\hat{m} = \begin{pmatrix} \hat{a} & \hat{b} \\ \bar{\hat{b}} & \bar{\hat{a}} \end{pmatrix} \quad \text{and} \quad m = \begin{pmatrix} a & b \\ c & d \end{pmatrix}$$

are related by $\hat{m} = lml^{-1}$, we have

$$\hat{a} = \tfrac{1}{2}(a + d + \mathrm{i}(b - c)), \qquad \hat{b} = \tfrac{1}{2}(a - d - \mathrm{i}(b + c)).$$

Hence the translation of our former conditions yields the following explicit result:

$$C_{n,r} = \left\{ m \in Sp(n, \mathbb{R}) | a = \begin{pmatrix} * & 0 \\ * & * \end{pmatrix}, c = \begin{pmatrix} * & 0 \\ 0 & 0 \end{pmatrix}, d = \begin{pmatrix} * & * \\ 0 & * \end{pmatrix} \right\},$$

$$B_{n,r} = \left\{ m \in Sp(n, \mathbb{R}) | a = \begin{pmatrix} \pm 1 & 0 \\ * & * \end{pmatrix}, b = \begin{pmatrix} 0 & * \\ * & * \end{pmatrix}, c = 0, d = \begin{pmatrix} \pm 1 & * \\ 0 & * \end{pmatrix} \right\}.$$

For any fixed r we consider two important projections

$$*: H_n \to H_r, \qquad z = \begin{pmatrix} z^* & * \\ * & * \end{pmatrix} \mapsto z^*,$$

$$*: C_{n,r} \to Sp(r, \mathbb{R}), \qquad m \mapsto m^* = \begin{pmatrix} a_1 & b_1 \\ c_1 & d_1 \end{pmatrix},$$

where $a_1, \ldots, d_1$ are the upper left $(r \times r)$-blocks of $a, \ldots, d$. Both will be denoted as projections onto the standard boundary component of degree r according to the interpretation above. These two mappings are immediately seen to be compatible in the sense that

$$m\langle z \rangle^* = m^*\langle z^* \rangle \tag{10}$$

holds for any $m \in C_{n,r}$, $z \in H_n$. Henceforth we shall be mainly interested in the intersections of the fixed groups with the modular group Γ_n, for which we use the same notation $C_{n,r}$ and $B_{n,r}$ as before.

Now we return to the question of how to lift cusp forms to modular forms of higher degree. The main instrument will be a certain type of Eisenstein series, by which modular forms can be lifted from a boundary component to all of H_n.

Definition 2

For $n \geq 1$, $0 \leq r \leq n$, even positive k and a cusp form $f \in S_r^k$ the Eisenstein series attached to f is the series

$$E_{n,r}^k(z; f) = \sum_{m \in C_{n,r} \backslash \Gamma_n} f(m\langle z\rangle^*) j(m, z)^{-k} \qquad (z \in H_n).$$

Let us consider two special cases. If $r = 0$, f is any complex number and $C_{n,0}$ is the group of all integral modular substitutions introduced in (3.3). Then the Eisenstein series of the definition coincides with the one introduced by C.L. Siegel in [62]. We write $E_{n,0}^k(z)$, if f is one. If $r = n$ we have $C_{n,n} = \Gamma_n$ and the Eisenstein series consists only of one term which is $f(z)$. Then we have to check whether Eisenstein series are well defined. Let $m_1 \in C_{n,r}$, $m \in \Gamma_n$, then by the transformation law of modular forms and because of (10) we obtain

$$f(m_1 m\langle z\rangle^*) = f(m_1^*\langle m\langle z\rangle^*\rangle) = j(m_1^*, m\langle z\rangle^*)^k f(m\langle z\rangle^*).$$

On the other hand, by the cocycle relation for Jacobians, and since k is even, we get

$$j(m_1 m, z)^{-k} = j(m_1, m\langle z\rangle)^{-k} j(m, z)^{-k} = j(m_1^*, m\langle z\rangle^*)^{-k} j(m, z)^{-k}.$$

Hence the general term of the Eisenstein series depends only on the left-coset of m modulo $C_{n,r}$.

Remark

There is no reasonable way to introduce Eisenstein series for odd weights.

Eisenstein series are equipped with excellent convergence properties. The modular form $f(z)$ of degree r, which appears in the definition of the Eisenstein series, is a cusp form and can be estimated by $\det y^{-k/2}$ according to Proposition 4. So we obtain the majorant

$$H_{n,r}^k(z) = \sum_{m \in C_{n,r} \backslash \Gamma_n} \det(\operatorname{Im} m\langle z\rangle^*)^{-k/2} |j(m, z)|^{-k} \tag{11}$$

uniformly in H_n. The series (11) can be majorized again by itself at a certain point, say $H_{n,r}^k$ (i1) uniformly in a vertical strip of positive height

$$V_n(d) = \{z \in H_n | \sigma(x^2) \leq d^{-1}, y \geq d\mathbf{1}\} \qquad (d > 0).$$

This is a consequence of the following

Lemma 2

For any $n \geq 1$, $0 \leq r \leq n$ and real $d > 0$ there exists a positive constant $\alpha = \alpha(n, d)$ depending only on n and d such that

$$\det(\operatorname{Im} m\langle z\rangle^*) |j(m, z)|^2 \geq \alpha \det(\operatorname{Im} m\langle i\mathbf{1}\rangle^*) |j(m, i\mathbf{1})|^2$$

for all $z \in V_n(d)$ and $m \in Sp(n, \mathbb{R})$.

Proof

Note that both sides of this inequality depend only on the left-coset of m modulo $C_{n,r}$. Using an appropriate element of the form (2) as left-hand factor of m we see that $m\langle z\rangle^*$ and $m\langle i\mathbf{1}\rangle^*$ may be submitted simultaneously to an arbitrary symplectic mapping. We take advantage of this freedom in the following way. It was proved in §1 (as corollary to a lemma) that two points of the generalized unit-circle can be mapped simultaneously into zero and a diagonal matrix t with $0 \le t < \mathbf{1}$. After transition to Siegel's half-space via Cayley's transformation this means that two arbitrary points of the half-space can be mapped simultaneously into $i\mathbf{1}$ and a diagonal matrix it, where $t \ge \mathbf{1}$. Hence it is sufficient to prove

$$|j(m, z)|^2 \ge \alpha |j(m, i\mathbf{1})|^2 \tag{12}$$

for all $z \in V_n(d)$, $m \in Sp(n, \mathbb{R})$. Note that the elements $m \in Sp(n, \mathbb{R})$ with non-singular lower left block c are dense in $Sp(n, \mathbb{R})$. We may therefore assume $\det c \ne 0$ in (12), and even $c = \mathbf{1}$. So we are left with the proof of the inequality

$$|\det(z + s)| \ge \alpha^{1/2} |\det(i\mathbf{1} + s)| \tag{13}$$

for all $z \in V_n(d)$ and all symmetric real matrices s. Set

$$y^{-1}\{\bar{z} + s\} = g^{-1}\,{}^t g^{-1}, \qquad y^{-1} = f\,{}^t f,$$

with $g, f \in GL(n, \mathbb{R})$ and introduce the real matrices p, q by

$$p = g(x + s)\,{}^t f, \qquad q = g f^{-1}.$$

Then a straightforward calculation yields

$$p\,{}^t p + q\,{}^t q = \mathbf{1}, \qquad q f(i\mathbf{1} - x)\,{}^t f + p = g(i\mathbf{1} + s)\,{}^t f.$$

From the first equation we deduce that p and q are bounded; since $z \in V_n(d)$ we know the boundedness of x and f. The last equality above then guarantees the boundedness of $g(i\mathbf{1} + s)\,{}^t f$, which implies (13).

So far we have shown that the Eisenstein series $E^k_{n,r}(z; f)$ converges absolutely and uniformly on any vertical strip of positive height provided the majorant (11) converges at a single point, for instance $z = i\mathbf{1}$. To prepare for the convergence proof for $H^k_{n,r}(i\mathbf{1})$ by methods borrowed from symplectic geometry we need

Lemma 3

To any compact subset K of H_n there exists a positive constant $\beta = \beta(n, K)$ depending only on n and K such that

$$\beta^{-1}\,\mathrm{Im}\, m\langle i\mathbf{1}\rangle \le \mathrm{Im}\, m\langle z\rangle \le \beta\,\mathrm{Im}\, m\langle i\mathbf{1}\rangle$$

for all $z \in K$ and $m \in Sp(n, \mathbb{R})$.

Proof
Take the inverse in the inequality stated above and use (1.6). Then the assertion can be reformulated as

$$\beta^{-1}\mathbf{1}\{{}^t(ci+d)\} \le y^{-1}\{{}^t(cz+d)\} \le \beta\mathbf{1}\{{}^t(ci+d)\}$$

for all $z \in K$ and $m \in Sp(n, \mathbb{R})$. We may again restrict ourselves to non-singular c, and after dividing by $|\det c|^2$ we are left with the proof of

$$\beta^{-1}\mathbf{1}\{i\mathbf{1}+s\} \le y^{-1}\{z+s\} \le \beta\mathbf{1}\{i\mathbf{1}+s\}$$

for all $z \in K$ and all symmetric real matrices s. Now

$$y^{-1}\{z+s\} = y^{-1}[x+s] + y,$$

and since $z \in K$ there exists a $\beta_1 = \beta_1(n, K)$ such that $\beta_1^{-1} \le y^{-1}$, $y \le \beta_1$. Hence we have

$$\beta_1^{-1}(\mathbf{1}+(x+s)^2) \le y^{-1}\{z+s\} \le \beta_1(\mathbf{1}+(x+s)^2).$$

On the other hand, since x is bounded,

$$\beta_2^{-1}(\mathbf{1}+s^2) \le \mathbf{1}+(x+s)^2 \le \beta_2(\mathbf{1}+s^2)$$

for all s with a similar constant $\beta_2 = \beta_2(n, K)$. The assertion is valid for $\beta = \beta_1\beta_2$.

Corollary
To any compact subset K of H_n there exists a constant $\gamma = \gamma(n, K)$ depending only on K and n such that

$$\operatorname{Im} m\langle z\rangle \le \gamma \operatorname{Im} m\langle \hat{z}\rangle$$

for all $z, \hat{z} \in K$ and all $m \in Sp(n, \mathbb{R})$.

Returning to the convergence of the majorant (11) at the point $z = i\mathbf{1}$, decompose $\operatorname{Im} m\langle z\rangle = \eta$ into

$$\eta = \begin{pmatrix} \eta_1 & 0 \\ 0 & \eta_2 \end{pmatrix}\left[\begin{pmatrix} \mathbf{1} & \eta_3 \\ 0 & \mathbf{1} \end{pmatrix}\right]$$

and write

$$H_{n,r}^k(i\mathbf{1}) = \sum_{m \in C_{n,r}\backslash\Gamma_n} \det{}^{k/2} \eta_2(i\mathbf{1}).$$

Choose any compact subset K of positive volume contained in Siegel's fundamental domain F_n. Then by Lemma 3 there exists a constant β_1 such that

$$\det{}^{k/2} \eta_2(i\mathbf{1}) \le \beta_1 \det{}^{k/2} \eta_2(z)$$

for all $z \in K$ and $m \in Sp(n, \mathbb{R})$. We integrate over K with respect to the symplectic metric on both sides of this inequality and obtain

$$\det{}^{k/2} \eta_2(i1) \leq \beta_2 \int_K \det{}^{k/2} \eta_2(z)\, dv_n$$
$$= \beta_2 \int_{m\langle K\rangle} \det{}^{k/2} y_2\, dv_n$$

with β_2 independent of m. Here dv_n denotes the symplectic volume element of degree n and

$$y = \begin{pmatrix} y_1 & 0 \\ 0 & y_2 \end{pmatrix}\left[\begin{pmatrix} 1 & y_3 \\ 0 & 1 \end{pmatrix}\right]$$

is decomposed in a similar way as η. Hence the convergence of $H^k_{n,r}(i1)$ is settled if we are able to prove

$$\int_{\bigcup m\langle K\rangle, m \in C_{n,r}\backslash\Gamma_n} \det{}^{k/2} y_2\, dv_n < \infty. \tag{14}$$

Note that the integrand is invariant with respect to $C_{n,r}$ and that the region of integration is part of a fundamental domain of $C_{n,r}$ acting on H_n. Therefore we may replace the region of integration by any fundamental domain Q of $C_{n,r}$ in H_n. Furthermore we may cut off a part of Q by the condition $\det y \leq a$ for appropriate a because the points in $K \subset F_n$ have this property, and beyond this they are of maximal height in their orbits. So to prove (14) we have to investigate the convergence of

$$\int_{Q, \det y \leq a} \det{}^{k/2} y_2\, dv_n.$$

It may be immediately checked that

$$Q = \{z \in H_n | z^* \in F_r,\ x, y_3 \text{ reduced mod } 1,\ y_2 \in R_{n-r}\} \tag{15}$$

is a fundamental domain of $C_{n,r}$ and

$$dv_n = \det y^{-n-1}\, dx\, dy, \qquad dy = \det y_1^{n-r}\, dy_1\, dy_2\, dy_3.$$

Hence we obtain for the integral in question

$$\int_{Q, \det y \leq a} \det y_1^{-r-1} \det y_2^{k/2-n-1}\, dx\, dy_1\, dy_2\, dy_3$$
$$\leq \int_{F_r} dv_r \int_{y_2 \in R_{n-r}, \det y_2 \leq b} \det y_2^{k/2-n-1}\, dy_2,$$

where we have carried out the trivial part of the integration. The condition $\det y_2 \leq b$ for appropriate b is explained as follows: $\det y = \det y_1 \det y_2$ is bounded from above by a and $\det y_1$ from below, since z^* is located in Siegel's fundamental domain. The integral over F_r converges, since F_r has finite symplectic volume. The remaining integral is finite for $k > n + r + 1$ by the following

Lemma 4

Let $n \geq 1$ and b be any positive real number. Then the integral

$$\int_{y \in R_n,\, \det y \leq b} \det y^s \, dy$$

converges for $s > -(n + 1)/2$.

We postpone the proof of this lemma in order to summarize our results.

Theorem 1

Let $n \geq 1, 0 \leq r \leq n, k > n + r + 1$ and even. For any cusp form f of degree r and weight k the Eisenstein series $E^k_{n,r}(z; f)$ converges absolutely and uniformly on any vertical strip of positive height.

For $r = 0$ we obtain the condition $k > n + 1$, first proved by H. Braun [12]. It is the exact region of absolute convergence in that case. Since locally we have uniform convergence, Eisenstein series represent holomorphic functions on H_n. The transformation law for a modular form of degree n and weight k follows from the cocycle relation for $j(m, z)^k$, the summation condition and the absolute convergence of $E^k_{n,r}(z; f)$. Hence Eisenstein series represent modular forms.

Proof (Lemma 4)

We use the notation of §2, in particular Definition 2.3. Enclose R_n within $Q'_n(a)$ and introduce Jacobian coordinates $y = d[v]$. A straightforward calculation yields

$$\det\left(\frac{\partial y}{\partial(d, v)}\right) = \prod_{k=1}^{n} d_k^{n-k}.$$

Then replace the variables $d_1, \ldots, d_n$ by

$$u_1 = \frac{d_1}{d_2}, \ldots, u_{n-1} = \frac{d_{n-1}}{d_n}, \qquad t = d_1 \ldots d_n;$$

the Jacobian of this transformation is $n\, d_1/d_n$. For the combined change of variables we obtain

$$\det\left(\frac{\partial y}{\partial(u, t, v)}\right) = n^{-1} \frac{d_n}{d_1} \prod_{k=1}^{n} d_k^{n-k} = n^{-1} t^{(n-1)/2} \prod_{v=1}^{n-1} u_v^{v(n-v)/2-1}.$$

So we are able to estimate the integral in question by

$$\int\limits_{\substack{y \in Q'_n(a) \\ \det y \leq b}} \det y^s \, dy = n^{-1} \int\limits_{\substack{0 \leq u_\nu \leq a \\ |v_{kl}| \leq a, 0 \leq t \leq b}} \prod_{\nu=1}^{n-1} u_\nu^{\nu(n-\nu)/2-1} t^{s+(n-1)/2} \, du \, dt \, dv.$$

Only the integral over t is critical. It is finite if the exponent of t satisfies $s + (n-1)/2 > -1$.

We are now able to recognize that the formation of Eisenstein series is an appropriate procedure to lift cusp forms with respect to Φ.

Proposition 5

Let $n \geq 1, 0 \leq r \leq n, k > n + r + 1$ and even. Then for any cusp form $f \in S_r^k$

$$E_{n,r}^k(*; f)|\Phi^{n-r} = f.$$

Corollary

Φ: $M_n^k \to M_{n-1}^k$ is surjective for even $k > 2n$.

Proof

The corollary is an immediate consequence of the proposition since then (5) is valid for any $0 \leq r \leq n$. Eisenstein series converge uniformly on vertical strips of positive height, therefore we may apply the operator Φ^{n-r} term by term. So we have to determine the limits (3),

$$\lim_{\lambda \to \infty} f(m\langle z\rangle^*) j(m, z)^{-k}, \qquad z = \begin{pmatrix} z^* & 0 \\ 0 & \mathrm{i}\lambda \mathbf{1} \end{pmatrix},$$

for any $m \in \Gamma_n$. The 'first' term of the Eisenstein series with $m \in C_{n,r}$ is just $f(z^*)$ and does not depend on λ. We will show that all the other terms tend to zero if $\lambda \to \infty$. In fact, this is valid for the majorant (11), i.e. we will prove

$$\lim_{\lambda \to \infty} \det(\operatorname{Im} m\langle z\rangle^*)^{-1} |j(m\langle z\rangle)|^{-2} = 0$$

for any $m \notin C_{n,r}$. For this purpose note that

$$\det(\operatorname{Im} m\langle z\rangle^*) |j(m, z)|^2 = \det y^* \, P(\lambda), \qquad z = \begin{pmatrix} z^* & 0 \\ 0 & \mathrm{i}\lambda \mathbf{1} \end{pmatrix}$$

with a polynomial $P(\lambda)$ in λ. Hence we only have to verify the degree of this polynomial to be positive. A straightforward calculation shows

$$P(\lambda) = \det(\lambda y^{*-1}\{{}^t(c_3 z^* + d_3)\} + \mathbf{1}\{{}^t(\mathrm{i}\lambda c_4 + d_4)\}),$$

where $(c_3 c_4 d_3 d_4)$ consists of the last $n - r$ rows of m. Now

$$\lambda y^{*-1}\{{}^t(c_3 z^* + d_3)\}, \qquad \mathbf{1}\{{}^t(\mathrm{i}\lambda c_4 + d_4)\}$$

are positive semi-definite, the sum being positive since $P(\lambda) \neq 0$. This observation yields

$$\det(\mathbf{1}\{{}^t(i\lambda c_4 + d_4)\}) \leq \det(\lambda y^{*-1}\{{}^t(c_3 z^* + d_3)\} + \mathbf{1}\{{}^t(i\lambda c_4 + d_4)\}) = P(\lambda).$$

Now assume $P(\lambda)$ to be constant. Then neither side of this inequality depends on λ and both have the common value $\det(d_4{}^t d_4) \neq 0$. We may infer

$$y^{*-1}\{{}^t(c_3 z^* + d_3)\} = 0$$

which implies $c_3 = d_3 = 0$, and then $c_4 = 0$. Hence $m \in C_{n,r}$, and the proposition is proved.

We add some comments on the surjectivity of Siegel's Φ-operator. The assumption that k is even in the corollary is essential. In fact, we know that $M_n^k = 0$, if k and n are odd. On the other hand, then $M_{n-1}^k \neq 0$ for sufficiently large k. Hence Φ cannot be surjective in that case. Concerning the condition $k > 2n$ further progress has been achieved recently by R. Weissauer [69]. He was able to consider the question of whether Φ is surjective, also for smaller values of k. Roughly speaking, Φ is shown to be surjective for even $k > n + 1$; but a complete answer is still awaited. Our condition $k > 2n$ was caused by the convergence of the Eisenstein series. Weissauer applies the so-called Hecke summation to $E_{n,r}^k$. Two phenomena make the situation for small values of k rather complicated as it may happen that the Hecke summation leads to non-holomorphic forms or to the zero-function. Moreover, the work of R.P. Langlands on the analytic continuation and the functional equation of Eisenstein series is involved in Weissauer's investigations. The problem of lifting a cusp form is related to the behavior of certain L-functions attached to the given form. In §8 we shall prove the surjectivity of Φ for $k < (n - 1)/2$ by another method.

From the surjectivity of the linear map

$$\Phi: M_n^k \to M_{n-1}^k$$

we deduce the isomorphism

$$M_{n-1}^k \simeq M_n^k / S_n^k$$

for even $k > 2n$. This formula is useful to obtain results on M_n^k from the subspaces of cusp forms by an induction argument. As an example any information on the dimension of M_n^k can be reduced to cusp forms in this way. In the next section we will present a new proof for the finiteness of $\dim S_n^k$, which is based on the theory of integral equations. Then we can immediately infer the corresponding result for M_n^k by the above argument.

Another consequence of Proposition 5 is concerned with the linear generation of M_n^k.

Proposition 6

For even $k > 2n$ the Eisenstein series

$$\{E^k_{n,r}(*;f) | f \in S^k_r, 0 \le r \le n\}$$

span the vector space of all modular forms of degree n and weight k.

Proof

Let f be any element of M^k_n. Then $f|\Phi^n \in S^k_0$. Determine the Eisenstein series $E^k_{n,0}$ by Proposition 5 such that $E^k_{n,0}|\Phi^n = f|\Phi^n$. Hence $(f - E^k_{n,0})|\Phi^{n-1}$ is a cusp form of degree one. In the next step take an Eisenstein series $E^k_{n,1}$ such that $E^k_{n,1}|\Phi^{n-1} = (f - E^k_{n,0})|\Phi^{n-1}$, hence $(f - E^k_{n,0} - E^k_{n,1})|\Phi^{n-2}$ becomes a cusp form of degree two. After n steps we obtain

$$f - E^k_{n,0} - \cdots - E^k_{n,n-1} \in S^k_n,$$

and any cusp form of degree n may be considered as an Eisenstein series of type $E^k_{n,n}$.

We would like to characterize Eisenstein series of different types by metrical properties. For this purpose we introduce the metrization integral of H. Petersson.

Definition 3

Let f, g be modular forms of weight k and degree n, at least one of which is a cusp form. The Petersson scalar product of f and g is the value of the integral

$$\{f, g\} = \int_{F_n} f(z)\overline{g(z)} \det y^k \, dv_n,$$

where F_n is Siegel's or any other measurable fundamental domain of Γ_n.

The importance of this metrization integral was recognized by H. Petersson in the one-variable case. It is an improper integral, the integrand of which is obviously invariant with respect to Γ_n. Moreover,

$$f(z)\overline{g(z)} \det y^k$$

is bounded by Proposition 4 since $f\bar{g}$ is a cusp form of weight $2k$, and F_n has finite symplectic volume. Hence the convergence of this improper integral is always guaranteed. The scalar product is a sesquilinear function. If restricted to

$$\{\ \}: S^k_n \times S^k_n \to \mathbb{C},$$

it defines a positive definite Hermitian form on S^k_n by which S^k_n becomes a unitary space.

In general we may decompose

$$M_n^k = S_n^k \oplus N_n^k \tag{16}$$

into S_n^k and its orthogonal complement N_n^k with respect to the scalar product. The action of Φ is trivial on S_n^k and injective on N_n^k. For if f is in N_n^k and mapped by Φ to zero it is a cusp form perpendicular to itself, hence it vanishes identically. We obtain a certain refinement of this decomposition if we introduce the following subspaces of M_n^k characterized by metrical properties [49].

Definition 4
Set $M_{0,0}^k := M_0^k$ and for $n \geq 1, 0 \leq r \leq n$

$$M_{n,r}^k := \begin{cases} S_n^k & \text{if } r = n \\ \{f \in N_n^k \mid f|\Phi \in M_{n-1,r}^k\} & \text{if } 0 \leq r < n. \end{cases}$$

Using the surjectivity of Φ for even $k > 2n$ we prove

Proposition 7
For $n \geq 0$ and even $k > 2n$ the vector space M_n^k of all modular forms of weight k and degree n is the direct sum

$$M_n^k = \sum_{r=0}^{n} \oplus M_{n,r}^k.$$

Proof
We argue by induction on n. Since Φ is surjective we have

$$\Phi(N_n^k) = M_{n-1}^k,$$

and we may assume as induction hypothesis

$$M_{n-1}^k = \sum_{r=0}^{n-1} \oplus M_{n-1,r}^k.$$

Now it can be deduced immediately from Definition 4 that $\Phi(M_{n,r}^k) = M_{n-1,r}^k$ for $0 \leq r < n$. So we obtain

$$\Phi(N_n^k) = \sum_{r=0}^{n-1} \oplus \Phi(M_{n,r}^k).$$

As Φ is injective on N_n^k we conclude that

$$N_n^k = \sum_{r=0}^{n-1} \oplus M_{n,r}^k,$$

which together with (16) proves the proposition.

The metrical characterization of different types of Eisenstein series can now be realized with the help of these subspaces $M^k_{n,r}$ and the corresponding decomposition of M^k_n.

Proposition 8

Let $n \geq 1, 0 \leq r \leq n, k > n + r + 1$ and even. Then

$$M^k_{n,r} = \{E^k_{n,r}(*; f) | f \in S^k_r\}.$$

Proof

(i) The operator Φ maps Eisenstein series onto Eisenstein series diminishing the degree by one and leaving all the other parameters fixed; more precisely, for $n \geq 1, 0 \leq r \leq n$, even $k > n + r + 1$ and any cusp form $f \in S^k_r$

$$E^k_{n,r}(*; f)|\Phi = \begin{cases} E^k_{n-1,r}(*; f) & \text{for } 0 \leq r < n \\ 0 & \text{for } r = n. \end{cases} \tag{17}$$

Here we have to define $E^k_{0,0}(*; f) := f$. Indeed (17) can be proved as follows. The cases $r = n$ or $n = 1$ are trivial or are contained in our former investigations. So we may assume $n > 1, r < n$. We apply the Φ-operator term by term and set

$$z = \begin{pmatrix} z_1 & 0 \\ 0 & i\lambda \end{pmatrix},$$

where z_1 is $(n - 1)$-rowed. Those terms of $E^k_{n,r}(z; f)$, for which m can be chosen in $C_{n,n-1}$, are independent of λ. Their totality amounts to

$$\sum_{m \in C_{n,r} \cap C_{n,n-1} \backslash C_{n,n-1}} f(m\langle z \rangle^*) j(m, z)^{-k} = E^k_{n-1,r}(z_1; f).$$

It can be shown that the other terms tend to zero if $\lambda \to \infty$ by a similar argument to that in Proposition 5. For details we refer to [44].

(ii) We prove

$$E^k_{n,r}(*; f) \in N^k_n$$

for $n \geq 1$, $0 \leq r < n$ and even $k > n + r + 1$. Let g be any cusp form of degree n and weight k and evaluate the Petersson scalar product of the Eisenstein series $E^k_{n,r}(*; f)$ with g:

$$\begin{aligned} \{E^k_{n,r}(*; f), g\} &= \int_{F_n} E^k_{n,r}(z; f)\overline{g(z)} \det y^k \, dv_n \\ &= \sum_{m \in C_{n,r} \backslash \Gamma_n} \int_{F_n} f(m\langle z \rangle^*) j(m, z)^{-k} \overline{g(z)} \det y^k \, dv_n \\ &= \sum_{m \in C_{n,r} \backslash \Gamma_n} \int_{m\langle F_n \rangle} f(z^*) \overline{g(z)} \det y^k \, dv_n \\ &= 2^{-\varepsilon_r} \int_{Q} f(z^*) \overline{g(z)} \det y^k \, dv_n, \end{aligned}$$

where Q denotes the fundamental domain of $C_{n,r}$ introduced in (15) and

$$\varepsilon_r = \begin{cases} 0 & \text{for } r = 0 \\ 1 & \text{for } r = 1,\dots,n-1. \end{cases}$$

To integrate term by term is legitimate, since Eisenstein series converge uniformly on Siegel's fundamental domain by Theorem 1 and $g(z) \det y^k$ is bounded according to Proposition 3. Moreover, the integrals over the single terms

$$\int_{F_n} f(m\langle z\rangle^*) j(m, z)^{-k} \overline{g(z)} \det y^k \, dv_n$$

exist, since, besides $g(z) \det y^k$, the term $f(m\langle z\rangle^*) j(m, z)^{-k}$ is bounded on F_n by Lemma 2. Note that $f(z^*)\overline{g(z)} \det y^k$ is invariant with respect to $C_{n,r}$, so we can take any fundamental domain of $C_{n,r}$ instead of Q. The factor $2^{-\varepsilon_r}$ is caused by the fact that Q is mapped onto itself by

$$z \mapsto z[u], \qquad u = \begin{pmatrix} 1 & 0 \\ 0 & \pm 1 \end{pmatrix}.$$

Now apply Fubini's theorem to the integral over Q. The integration over x_4, the lower right block of size $n - r$ in x, leads to

$$\int_{x_4 \bmod 1} \overline{g(z)} \, dx_4.$$

But this integral already vanishes, as can be seen from the Fourier series of g being a cusp form. Hence $E^k_{n,r}$ is orthogonal to all cusp forms of weight k and degree n.

(iii) By induction on n we infer from (i) and (ii) that

$$E^k_{n,r}(*; f) \in M^k_{n,r}$$

for $n = 0, 1, 2, \dots, 0 \le r \le n$ and even $k > n + r + 1$. So, using Proposition 5, the mapping

$$\Phi^{n-r}: M^k_{n,r} \to S^k_r$$

is surjective. Moreover it is injective, since $M^k_{n,r} \subset N^k_n$ for $r < n$. The statement of the proposition follows from the correspondence between $E^k_{n,r}(*; f)$ and f under this isomorphism.

We summarize the main results of this section in

Theorem 2

Let $n \ge 0$, $k > 2n$ and even. The spaces M^k_ν of modular forms of common weight k and degrees $\nu = 0, 1, \dots, n$ decompose into the direct sum of *metrically characterized subspaces*

$$\begin{array}{ccccccccc}
M_n^k & = M_{n,0}^k & \oplus M_{n,1}^k & \oplus \cdots \oplus M_{n,n-1}^k & \oplus M_{n,n}^k \\
 & \downarrow & \downarrow & \downarrow & \\
M_{n-1}^k & = M_{n-1,0}^k & \oplus M_{n-1,1}^k & \oplus \cdots \oplus M_{n-1,n-1}^k & \\
\vdots & \vdots & \vdots & & \\
 & \downarrow & \downarrow & & \\
M_1^k & = M_{1,0}^k & \oplus M_{1,1}^k & & \\
 & \downarrow & & & \\
M_0^k & = M_{0,0}^k. & & &
\end{array}$$

Each subspace $M_{n,r}^k$ consists of the Eisenstein series $E_{n,r}^k$ indexed in the same manner. Φ operates bijectively, where the arrows indicate it. The inverse map $\psi_{n,r}$ of

$$\Phi^{n-r}: M_{n,r}^k \to M_{r,r}^k$$

assigns to each cusp form $f \in S_r^k$ the Eisenstein series

$$\psi_{n,r}(f) = E_{n,r}^k(*; f).$$

III

Large weights

6 Cusp forms and Poincaré series

In the previous section we saw why it is important to concentrate on cusp forms. Now we will examine the space S_n^k of cusp forms more closely. We look on S_n^k as a unitary space equipped with the scalar product introduced in Definition 5.3 and we shall first study Petersson's metrization theory in more detail. In particular we will be able to generate S_n^k for large values of k by different types of Poincaré series. These different types will turn out to be important for different purposes; nevertheless they are related in a non-trivial manner. Roughly speaking, cusp forms cover exactly that part of modular forms which can be investigated by methods borrowed from the theory of automorphic functions on bounded domains. Hence it is advisable (although avoidable), in order to make our arguments more transparent, to consider besides Siegel's upper half-plane H_n the bounded realization D_n.

Cayley's transformation was introduced in Proposition 1.2. It maps H_n onto D_n and was normed in such a way that $\mathrm{i}\mathbf{1}$ is mapped into zero. Of course we can replace $\mathrm{i}\mathbf{1}$ by any other point $w = u + \mathrm{i}v$ in H_n; then we get the following generalization. Determine for given w a real n-rowed matrix q of positive determinant such that $v[{}^tq] = \mathbf{1}$; consider the symplectic map

$$z \mapsto (z - u)[{}^tq],$$

which carries w into $\mathrm{i}\mathbf{1}$, followed by Cayley's transformation of Proposition 1.2. The composed map has the form

$$l_w\colon H_n \to D_n, \qquad z \mapsto \zeta = l_w\langle z\rangle = q(z - w)(z - \bar{w})^{-1}q^{-1} \tag{1}$$

and maps w into zero. We combine our former results on $l_{\mathrm{i}\mathbf{1}}$ with those on symplectic transformations to obtain the corresponding generalizations for l_w. Let us mention a few formulas explicitly for subsequent use. We derive from the determination of Jacobians in §1 and §3

$$\det\left(\frac{\partial l_w\langle z\rangle}{\partial z}\right) = (2\mathrm{i})^{n(n+1)/2}\det(z - \bar{w})^{-n-1}\det v^{(n+1)/2}.$$

If f is any modular form of weight k, then

$$\hat{f}(\zeta) = \det(z - \bar{w})^k f(z) \tag{2}$$

becomes an automorphic form with respect to the transformed group

$$\hat{\Gamma}_n = l_w \Gamma_n l_w^{-1}$$

acting on D_n. The factors of automorphy are

$$\frac{\det(m\langle z\rangle - \bar{w})^k \det(cz+d)^k}{\det(z-\bar{w})^k} = \det(\bar{\hat{b}}\zeta + \bar{\hat{a}})^k, \tag{3}$$

where

$$\hat{m} = \begin{pmatrix} \hat{a} & \hat{b} \\ \bar{\hat{b}} & \bar{\hat{a}} \end{pmatrix} = l_w m l_w^{-1}.$$

Finally we mention the formula

$$(1 - \bar{\zeta}\zeta)[q] = 4y\{(z-\bar{w})^{-1}\}, \tag{4}$$

which was proved in connection with Proposition 1.2 for $w = i1$ and is generalized to arbitrary w immediately.

We start our investigations with a reproducing formula for holomorphic functions on H_n, which is essentially due to A. Selberg.

Proposition 1

Assume $k > 2n$ and let f be any holomorphic function on H_n such that $\det y^{k/2} f(z)$ is bounded. Then f can be represented by

$$f(w) = a_{nk} \int_{H_n} f(z) \det(\bar{z} - w)^{-k} \det y^k \, dv_z,$$

where a_{nk} is a numerical constant depending only on n and k, and dv_z denotes the symplectic volume element with respect to z.

Proof

We transform the integral in question to D_n by Cayley's transformation (1). Because of (2) and (4) we obtain

$$\begin{aligned} &\int_{H_n} f(z)\det(\bar{z}-w)^{-k}\det y^k \, dv_z \\ &\qquad = 2^{n(n-2k+1)} \det v^{-k} \int_{D_n} \hat{f}(\zeta) \det(1 - \bar{\zeta}\zeta)^k \, dv_\zeta, \end{aligned} \tag{5}$$

where dv_ζ denotes the invariant volume element

$$dv_\zeta = \frac{d\xi \, d\eta}{\det(1 - \bar{\zeta}\zeta)^{n+1}} \qquad (\zeta = \xi + i\eta).$$

Concerning the convergence, note that the quantity $\det(1 - \bar{\zeta}\zeta)^{k/2} \hat{f}(\zeta)$ is essentially $\det y^{k/2} f(z)$ according to (4) and therefore bounded. Hence the convergence of our integral is reduced to the question of whether

$$\int_{D_n} \det(\mathbf{1} - \bar{\zeta}\zeta)^{k/2}\, dv_\zeta$$

is finite. This latter integral was studied by L.K. Hua in [31]. He was even able to compute its value; here we only take for granted that we have convergence for $k > 2n$, which is not difficult to verify. Now we expand $\hat{f}(\zeta)$ into a power series with center at 0. But D_n is no Reinhardt-domain. Hence we may not expect the power series to converge everywhere in D_n. But if one first collects the homogeneous terms of fixed degree ν to $\hat{f}_\nu(\zeta)$, then

$$\hat{f}(\zeta) = \sum_{\nu=0}^{\infty} \hat{f}_\nu(\zeta) \tag{6}$$

converges everywhere in D_n. Indeed, for any fixed $\zeta \in D_n$ consider the function

$$g(t) = \hat{f}(t\zeta)$$

of the single complex variable t in $|t| \leq 1$. Compare the Taylor expansion of g which holds in $|t| \leq 1$ with the Taylor expansion of $\hat{f}(\zeta)$ in all the variables ζ_{kl} ($k \leq l$) in a sufficiently small neighborhood of the origin. Then we find

$$g(t) = \sum_{\nu=0}^{\infty} t^\nu \hat{f}_\nu(\zeta) \qquad (|t| \leq 1)$$

and in particular (6) for $t = 1$. The expansion (6) is uniformly convergent on compact subsets of D_n. Hence we may evaluate (5), integrating term by term over

$$D_n^{(t)} = \{\zeta \mid t\mathbf{1} - \bar{\zeta}\zeta \geq 0\} \qquad (0 < t < 1)$$

and then passing to the limit $t \to 1$. Now

$$\int_{D_n^{(t)}} \hat{f}_\nu(\zeta) \det(\mathbf{1} - \bar{\zeta}\zeta)^k\, dv_\zeta$$

vanishes for $\nu > 0$, as can be seen by substituting $\zeta \mapsto e^{is}\zeta$ with real s. Thus we are left with the constant term, and using (2) we obtain

$$\begin{aligned}
&\int_{H_n} f(z) \det(\bar{z} - w)^{-k} \det y^k\, dv_z \\
&= 2^{n(n-2k+1)} \det v^{-k} \int_{D_n} \det(\mathbf{1} - \bar{\zeta}\zeta)^k\, dv_\zeta\, \hat{f}(0) \\
&= 2^{n(n-k+1)} i^{nk} \int_{D_n} \det(\mathbf{1} - \bar{\zeta}\zeta)^k\, dv_\zeta\, f(w) \\
&= a_{nk}^{-1} f(w).
\end{aligned}$$

Remark

The calculation of Hua's integral in [31] yields the following value for a_{nk},

$$a_{nk} = 2^{-n(n/2+3/2-k)} i^{-nk} \pi^{-n(n+1)/2} \prod_{\nu=1}^{n} \frac{\Gamma\left(k - \frac{\nu-1}{2}\right)}{\Gamma\left(k - \frac{n+\nu}{2}\right)}.$$

Corollary

Let $k > n$; then

$$\det v^{-k} = \alpha_{nk} \int_{H_n} \det y^k |\det(z - \overline{w})|^{-2k} \, dv_z,$$

the constant α_{nk} depending only on n and k.

The corollary, which is a kind of reciprocity formula for det y^k, follows from the proposition for $f(z) = \det(z - \overline{w})^{-k}$ if $k > 2n$. It can also be derived directly from (4) and the convergence of Hua's integral; this latter argument works for $k > n$. Finally we mention that the kernel in the reproducing formula of the proposition is essentially Bergman's kernel function if transformed to the bounded domain D_n.

From the proposition we deduce an interesting formula for arbitrary cusp forms f. Note that $\det y^{k/2} f(z)$ is bounded for any $f \in S_n^k$ by Proposition 5.4. Then formally we obtain for $k > 2n$, $kn \equiv 0 \bmod 2$ and $f \in S_n^k$

$$\begin{aligned} f(w) &= a_{nk} \int_{H_n} f(z) \det(\overline{z} - w)^{-k} \det y^k \, dv_z \\ &= a_{nk} \sum_{m \in \pm 1 \backslash \Gamma_n} \int_{F_n} f(m\langle z\rangle) \overline{\det(m\langle z\rangle - \overline{w})}^{-k} \det(\operatorname{Im} m\langle z\rangle)^k \, dv_z \\ &= a_{nk} \sum_{m \in \pm 1 \backslash \Gamma_n} \int_{F_n} f(z) \overline{\det(m\langle z\rangle - \overline{w})}^{-k} \overline{j(m, z)}^{-k} \det y^k \, dv_z, \end{aligned}$$

and finally

$$f(w) = a_{nk} \{ f, P_n^k(*, -\overline{w}) \}, \tag{7}$$

where we have denoted by

$$P_n^k(z, w) = \sum_{m \in \pm 1 \backslash \Gamma_n} \det(m\langle z\rangle + w)^{-k} j(m, z)^{-k} \tag{8}$$

the so-called Poincaré series of type I. Provided our procedure was legitimate we have deduced an integral equation (7) for arbitrary cusp forms, in which Petersson's metrization integral and the Poincaré series of type I are involved. This type of Poincaré series turns out to be of a well-known nature. If one transforms $P_n^k(z, -\bar{w})$ into D_n by Cayley's transformation (1), then

$$\det(z - \bar{w})^k P_n^k(z, -\bar{w})$$

considered as a function of z is just the usual Poincaré series of $\hat{\Gamma}_n$ formed with the powers of the Jacobians. Indeed, to see this one only has to consider (3). From Poincaré's theorem (cf. Theorem 3.3) we obtain the condition $k \geq 2n + 2$ for absolute and uniform convergence on compact subsets.

By studying the Poincaré series in Siegel's half-space, properties that are much more remarkable can be derived.

Proposition 2

Let k, n be integers, $k > 2n$ and kn even. Then the Poincaré series are absolutely convergent and satisfy

(i) $P_n^k(z, w) = P_n^k(w, z)$ *for* $z, w \in H_n$;

(ii) *after taking absolute values of its terms (denoted by* $\lceil\ \rceil$) the estimate

$$\lceil P_n^k(z, w) \rceil \prec \lceil P_n^k(i\mathbf{1}, i\mathbf{1}) \rceil$$

in the sense of a majorant holds uniformly in z, w in any product of two vertical strips $V_n(d) \times V_n(d)$ of positive height d, in particular we have uniform convergence in such big regions;

(iii) $\det y^{k/2} \lceil P_n^k(z, w) \rceil$ *is bounded uniformly in $z \in H_n$ and w in any compact subset of H_n (Godement's theorem).*

Corollary

$P_n^k(*, w)$ as well as $P_n^k(z, *)$ represent cusp forms of weight k and degree n.

The corollary is an immediate consequence of the proposition. Indeed we have uniform convergence on vertical strips of positive height by (ii), hence Poincaré series represent holomorphic functions. The transformation law for modular forms with respect to z can be read off from (8). Finally, apply Siegel's Φ-operator to (iii) in order to show that $P_n^k(*, w)$ is a cusp form. Using (i) we get similar results for $P_n^k(z, *)$.

Proof

Concerning (i) we easily verify the important but elementary symmetry relation

$$\det(m\langle z \rangle + w) j(m, z) = \det(\tilde{m}\langle w \rangle + z) j(\tilde{m}, w), \tag{9}$$

where

$$\tilde{m} = m^{-1}\left[\begin{pmatrix} 1 & 0 \\ 0 & -1 \end{pmatrix}\right].$$

Note that '$\sim$' describes an anti-automorphism of Γ_n and anticipate the absolute convergence of the Poincaré series at this moment.
In order to prove (ii) we need an estimate

$$|\det(z + w)| \geq \alpha |\det(i\mathbf{1} + w)| \tag{10}$$

uniformly in $z \in V_n(d)$ and $w \in H_n$ with an appropriate constant α only depending on n and d. We have already proved a similar estimate in (5.13), namely

$$|\det(z + s)| \geq \beta |\det(i\mathbf{1} + s)|$$

for all $z \in V_n(d)$ and all real symmetric s. The inequality (10) can be reduced to (5.13) by transforming $\mathbf{1} + v$ into the identity matrix. Now apply (10) twice to the general term of the Poincaré series, where the change from z to w has to be combined with the symmetry relation (9).
Godement's theorem will be proved by an argument that is essentially due to C.J. Earle [16]. Consider for $z, \hat{w} \in H_n$ the function

$$G_k(z, \hat{w}) = \int_{F_n} |\det(z + \hat{w})|^k |\det(z - \bar{w})|^{-k} \det v^{k/2} \, dv_w,$$

where F_n is any fundamental domain of Γ_n. Compare this integral with the integral in the corollary of Proposition 1 in order to verify its convergence for $k > 2n$. First we remark that $G_k(z, \hat{w})$ is bounded from below by a positive number $\alpha(K)$ if z varies in H_n and $\hat{w}$ in a compact subset K. Indeed one obtains a lower bound in the following manner. Replace F_n by a compact subset of positive volume and use the fact that the integrand is bounded from below by a positive number uniformly in $z \in H_n$ and w, $\hat{w}$ in any compact subset. Now for $m \in \Gamma_n$ we have by the definition of $G_k(z, \hat{w})$ and the symmetry relation (9)

$$\begin{aligned} &G_k(m\langle z\rangle, \hat{w}) |\det(m\langle z\rangle + \hat{w})|^{-k} |j(m, z)|^{-k} \\ &\quad = \int_{F_n} |\det(m\langle z\rangle - \bar{w})|^{-k} |j(m, z)|^{-k} \det v^{k/2} \, dv_w \\ &\quad = \int_{F_n} |\det(z + \tilde{m}\langle -\bar{w}\rangle)|^{-k} |j(\tilde{m}, -\bar{w})|^{-k} \det v^{k/2} \, dv_w \\ &\quad = \int_{F_n} |\det(z - \overline{m^{-1}\langle w\rangle})|^{-k} |j(m^{-1}, w)|^{-k} \det v^{k/2} \, dv_w. \end{aligned}$$

Take the sum over $m \in \pm 1\backslash\Gamma_n$ on both sides and apply the reciprocity formula of the corollary to Proposition 1. Then we obtain

$$\alpha(K)\,\overline{|P_n^k(z,\hat{w})|} \leq \int_{H_n} |\det(z-\overline{w})|^{-k} \det v^{k/2}\, dv_w$$
$$= \alpha_{nk}^{-1} \det y^{-k/2}$$

for arbitrary $z \in H_n$, $\hat{w} \in K$, which proves the last assertion of Proposition 2.

We know from part (ii) of Proposition 2 that $\overline{|P_n^k(z,w)|}$ converges uniformly in z and w on any product of two vertical strips of positive height provided $k > 2n$. In particular we have uniform convergence on compact subsets of $H_n \times H_n$ or on $F_n \times F_n$, where F_n is Siegel's fundamental domain. Concerning convergence the situation seems to be perfect. Nevertheless we want to point out that Godement's theorem (iii) alone is a rich source of convergence theorems if one variable, say w, is assumed to be fixed. For instance, one may infer from (iii) that the partial sums of $\overline{|P_n^k(*,w)|}$ are locally bounded; hence by Vitali's theorem we have uniform convergence of $P_n^k(z,w)$ with respect to z on any compact subset of H_n. The uniform convergence of $P_n^k(z,w)$ with respect to z on a vertical strip of positive height can then be derived from (iii) by the following argument. If $\det y$ is large the whole series $\overline{|P_n^k(z,w)|}$ becomes small by Godement's theorem; on the other hand the condition $\det y \leq t$ selects a compact part of a vertical strip of positive height. Hence we have uniform convergence of $P_n^k(z,w)$ with respect to z everywhere in the vertical strip. Finally we remark that (iii) and its proof hold for arbitrary discontinuous subgroups of $Sp(n,\mathbb{R})$.

A gap remained in the proof of the integral equation (7) for cusp forms which we are now able to close. It was concerned with the interchangeability of summation and integration. This procedure is legitimate because the relevant series is the Poincaré series $P_n^k(*,\overline{w})$, which converges uniformly on Siegel's fundamental domain, multiplied by the function $f(z)\det y^k$ which is bounded by Proposition 5.4. Having now proved (7) completely let us state

Theorem 1

Assume $k > 2n$ and $kn \equiv 0 \bmod 2$. Then any cusp form f of weight k and degree n satisfies

$$f(w) = a_{nk} \int_{F_n} f(z)\,\overline{P_n^k(z,-\overline{w})}\,\det y^k\, dv_z,$$

and, vice versa, any continuous solution of this integral equation with appropriate behavior at infinity is a cusp form.

The second part of the theorem is obvious, since $\overline{P_n^k(z, -\bar{w})}$ is a cusp form with respect to w. In fact we have proved $P_n^k(z, *)$ to be a cusp form, and the assignment

$$\#: f(w) \mapsto f^\#(w) := \overline{f(-\bar{w})}$$

always maps modular forms into modular forms, since

$$m \mapsto m\left[\begin{pmatrix} 1 & 0 \\ 0 & -1 \end{pmatrix}\right]$$

is an automorphism of Γ_n. The unspecified 'appropriate behavior at infinity' shall guarantee sufficiently smooth convergence properties of the integral.

The formula in the theorem is the starting point for determining the dimension of the space S_n^k of cusp forms by Selberg's trace-formula. Indeed, if $f_1, \ldots, f_d$ is an orthonormal basis of S_n^k, the integral equation shows that

$$\sum_j f_j(z)\overline{f_j(-\bar{w})}$$

is a representation of the kernel $P_n^k(z, w)$ up to a constant factor. Hence the dimension of S_n^k may be determined by evaluating the integral

$$\int_{F_n} P_n^k(z, -\bar{z}) \det y^k \, dv_z.$$

Here we only take the opportunity of presenting another proof of the finiteness of dim M_n^k. Without loss of generality we may restrict ourselves to cusp forms and may assume $k > 2n$ and kn to be even. So we have to prove that the integral equation of Theorem 1 has only finitely many linearly independent solutions. We reformulate this integral equation as

$$F(w) = a_{nk} \int_{F_n} F(z) \overline{K(z, w)} \, dv_z, \tag{11}$$

where we have set

$$K(z, w) = P_n^k(z, -\bar{w}) \det y^{k/2} \det v^{k/2}$$

and $F(z) = \det y^{k/2} f(z)$. Because of the symmetry relation (9), the new kernel becomes Hermitian,

$$K(z, w) = \overline{K(w, z)},$$

and the integral equation (11) appears in the customary form. Now it is well known from integral equations that the vector space of all (continuous) solutions is of finite dimension if the kernel $K(z, w)$ is of Hilbert–Schmidt type, i.e. if

$$\int_{F_n \times F_n} |K(z, w)|^2 \, dv_z \, dv_w < \infty.$$

In our case the condition is satisfied since the fundamental domain F_n has finite symplectic volume and $K(z, w)$ is bounded according to the following

Lemma 1

For $k > 2n$, $kn \equiv 0 \bmod 2$ *the function*

$$P_n^k(z, w) \det y^{k/2} \det v^{k/2}$$

is bounded everywhere in $H_n \times H_n$.

Proof

Since $P_n^k(z, w)$ is a cusp form in z and a cusp form in w by the corollary of Proposition 2, we have a Fourier expansion

$$P_n^k(z, w) = \sum_{s,t>0} b(s, t) e^{2\pi i \sigma(tz+sw)},$$

where t and s run over all positive definite half-integral matrices. First we prove the following estimate similar to Proposition 5.3. To any $c > 0$ there exist two positive numbers c_1, c_2 such that

$$|P_n^k(z, w)| \leq c_1 e^{-c_2\sigma(y+v)} \tag{12}$$

for all $z, w \in H_n$, for which y, v are reduced in the sense of Minkowski and $y, v \geq c\mathbf{1}$. Indeed, we can estimate the exponential term in two different ways. On the one hand we have

$$\sigma(ty + sv) \geq c\sigma(t + s)$$

since $y, v \geq c\mathbf{1}$; on the other we have

$$\sigma(ty + sv) \geq c_3\sigma(y + v)$$

since y, v are reduced. Hence we obtain

$$\begin{aligned} |P_n^k(z, w)| &\leq \sum_{s,t>0} |b(s, t)| e^{-\pi\sigma(ty+sv)} e^{-\pi\sigma(ty+sv)} \\ &\leq e^{-\pi c_3\sigma(y+v)} \sum_{s,t>0} |b(s, t)| e^{-\pi c\sigma(t+s)} \\ &\leq c_1 e^{-c_2\sigma(y+v)}. \end{aligned}$$

Now we argue as in the proof of Proposition 5.4. For given $z, w \in H_n$ determine modulo Γ_n equivalent points z^*, w^* in Siegel's fundamental domain. Then (12) may be applied to z^*, w^* and we obtain

$$\begin{aligned} |P_n^k(z, w)| \det y^{k/2} \det v^{k/2} &= |P_n^k(z^*, w^*)| \det y^{*k/2} \det v^{*k/2} \\ &\leq c_1 \det y^{*k/2} \det v^{*k/2} e^{-c_2\sigma(y^*+v^*)}. \end{aligned}$$

By the inequality of the geometric and arithmetic means

$$\det y^{*1/n} \leq \frac{\sigma(y^*)}{n}, \qquad \det v^{*1/n} \leq \frac{\sigma(v^*)}{n},$$

we infer

$$|P_n^k(z, w)| \det y^{k/2} \det v^{k/2} \leq c_1 \det y^{*k/2} \det v^{*k/2} e^{-c_2 n(\det y^{*1/n} + \det v^{*1/n})}$$

$$\leq a_1 e^{-a_2(\det y^{*1/n} + \det v^{*1/n})}$$

with constants a_1, a_2 independent of z, w. Hence $|P_n^k(z, w)| \det y^{k/2} \det v^{k/2}$ is bounded throughout $H_n \times H_n$.

We may look upon Theorem 1 as the main-formula of metrization theory for Poincaré series of type I. This notation was introduced by H. Petersson in the one-variable case and is perhaps justified by the following

Corollary

Let k, n be integers, $k > 2n$ and kn even.

(i) *$P_n^k(*, w)$ is orthogonal to any cusp form f which vanishes at $-\bar{w}$, and is determined uniquely by this condition up to a constant factor (metrical characterization of Poincaré series).*

(ii) *There exist finitely many points $w_1, \ldots, w_s \in H_n$ such that the Poincaré series $P_n^k(*, w_\nu)$ $(\nu = 1, \ldots, s)$ span the space S_n^k of cusp forms (completeness theorem).*

Both statements are immediate consequences of Theorem 1 and from linear algebra in the unitary space S_n^k. To verify for instance (ii), decompose S_n^k into the subspace spanned by all the Poincaré series $\{P_n^k(*, w) | w \in H_n\}$ and its orthogonal complement. Then any f in the orthogonal complement vanishes everywhere in H_n as can be seen immediately from the main-formula. Hence S_n^k is spanned by the Poincaré series of type I.

Remark

Nothing is known about how to determine points $w_1, \ldots, w_s$ explicitly such that (ii) is valid.

In the literature there appear many kinds of different Poincaré series. We try to reduce all of them to a general principle. Let (h_ν) be a countable family of holomorphic functions in H_n such that any modular form of degree n can be expanded into a series

$$f(z) = \sum_\nu a_\nu h_\nu(z), \tag{13}$$

and assume that the coefficients may be obtained from f by an integral transformation of the type

$$a_\nu = \int_K f(z)s_\nu(z)\,dz, \tag{14}$$

where $K \subset H_n$ is compact and s_ν is continuous on K. Then we call (h_ν) a type for the expansion of modular forms and (13) the (h_ν)-expansion of f. For instance, Fourier or Taylor expansions are of this kind. One has to take the elementary formula for the Fourier coefficients, respectively Cauchy's integral formula, as the integral transformation (14). Consider more generally than before the two mappings

$$\#: \mathrm{Hol}(H_n) \to \mathrm{Hol}(H_n), \qquad f(z) \mapsto \overline{f(-\bar{z})},$$

$$\sim: Sp(n,\mathbb{R}) \to Sp(n,\mathbb{R}), \qquad m \mapsto m^{-1}\left[\begin{pmatrix}1 & 0\\ 0 & -1\end{pmatrix}\right].$$

The first one preserves holomorphy; the second one is an anti-automorphism for groups. It is easily verified that f is an automorphic form with respect to a subgroup G of $Sp(n,\mathbb{R})$ if and only if $f^\#$ is an automorphic form with respect to $\tilde{G}$. In particular, the mapping $\#$ is an isomorphism for modular forms since $\tilde{\Gamma}_n = \Gamma_n$, and $\#$ sends any type (h_ν) for the expansion of modular forms into another such type $(h_\nu^\#)$, as can be seen immediately from the definitions.

Proposition 3

Let $k > 2n$ and kn be even; let Λ_n be a subgroup of Γ_n containing ± 1 and let (h_ν) be any type for the expansion of modular forms. Assume an expansion

$$\sum_{m \in \pm 1\backslash\Lambda_n} \det(m\langle z\rangle + w)^{-k} j(m,z)^{-k} = \sum_\nu g_\nu(z) h_\nu^\#(w), \tag{15}$$

where the coefficients $g_\nu(z)$ can again be determined by the integral transformation (14) *for the type $(h_\nu^\#)$. Then the Poincaré series*

$$G_n^k(z; g_\nu) = \sum_{m \in \Lambda_n\backslash\Gamma_n} g_\nu(m\langle z\rangle) j(m,z)^{-k}$$

represent modular forms of weight k and satisfy:

(i) *$G_n^k(z; g_\nu)$ converges absolutely and uniformly in z on any vertical strip of positive height,*

(ii) *$\det y^{k/2}\,|G_n^k(z; g_\nu)|$ is bounded on H_n (Godement's theorem),*

(iii) *$P_n^k(z,w) = \sum_\nu G_n^k(z; g_\nu) h_\nu^\#(w)$.*

Proof

First we remark that the subseries of $P_n^k(z, w)$ in (15) represents an automorphic form in w with respect to $\tilde{\Lambda}_n$ because of the symmetry relation (9). Therefore an $(h_\nu^\#)$-expansion can be expected but is by no means certain. The uniform convergence of $\boxed{G_n^k(z;g_\nu)}$ in vertical strips of positive height is a direct consequence of the uniform convergence of $\boxed{P_n^k(z, w)}$ (z in a vertical strip of positive height, w in a compact set), since the validity of Cauchy's criterion is carried over from $\boxed{P_n^k(z, w)}$ to $\boxed{G_n^k(z;g_\nu)}$ by the integral transformation (14) for $(h_\nu^\#)$. Godement's theorem can be transferred from $P_n^k(z, w)$ (cf. Proposition 2) in the same manner. Finally, $P_n^k(z, w)$ is a modular form with respect to w and its $(h_\nu^\#)$-expansion, if calculated by the integral transformation (14) for $(h_\nu^\#)$, turns out to be of the indicated form.

The principle of transferring properties of $P_n^k(z, w)$ to $G_n^k(z;g_\nu)$ via the integral transformation (14) was applied several times in the course of the proof. It may be extended to the corollary of Proposition 2 as well. So we obtain from there and statement (iii) of Proposition 3 the

Corollary

The Poincaré series $G_n^k(z;g_\nu)$ represent cusp forms of weight k and degree n.

Further efforts to apply our principle will be concerned with the main-formula of metrization theory and its consequences. Let $k > 2n$ and kn be even. Consider the (h_ν)-expansion of any cusp form f of weight k and degree n

$$f(w) = \sum_\nu a_\nu(f)h_\nu(w).$$

Then we obtain from (14), Theorem 1 and statement (iii) of Proposition 3

$$\begin{aligned} a_\nu(f) &= \int_K f(w)s_\nu(w)\,dw \\ &= a_{nk}\int_K\left(\int_{F_n} f(z)\,\overline{P_n^k(z, -\overline{w})}\det y^k\,dv_z\right)s_\nu(w)\,dw \\ &= a_{nk}\int_{F_n} f(z)\left(\int_K \overline{P_n^k(z, -\overline{w})\,s_\nu(w)\,dw}\right)\det y^k\,dv_z \\ &= a_{nk}\{f, G_n^k(*;g_\nu)\}. \end{aligned} \tag{16}$$

The application of Fubini's theorem is permitted since the integrand is bounded. Formula (16) may be considered as the main-formula of metrization theory for the Poincaré series $G_n^k(*;g_\nu)$. Then, analogously to Theorem 1, we have

$$f(w) = a_{nk} \sum_{\nu} \{f, G_n^k(*; g_\nu)\} h_\nu(w).$$

We can infer a completeness theorem by the well-known argument: if a cusp form is orthogonal to all the Poincaré series $G_n^k(*; g_\nu)$ then it vanishes identically. Note that in (16) two different types for the expansion of functions are involved. The $a_\nu(f)$ are the coefficients of the (h_ν)-expansion of f, and the Poincaré series $G_n^k(*; g_\nu)$ are formed with functions g_ν coming from (g_ν)-expansions. Both types, (h_ν) and (g_ν), are related by (15) and will be called complementary to each other. So we may state

Theorem 2

Under the same assumptions as in Proposition 3,

(i) *the coefficients $a_\nu(f)$ of the (h_ν)-expansion of any cusp form f are connected with the Poincaré series $G_n^k(*; g_\nu)$ formed with the complementary type (g_ν) by the metrization formula*

$$a_\nu(f) = a_{nk}\{f, G_n^k(*; g_\nu)\};$$

(ii) *there are finitely many g_ν such that the Poincaré series $G_n^k(*; g_\nu)$ span the space of cusp forms (completeness theorem).*

Proposition 3 offers a method of finding a large variety of Poincaré series and guarantees excellent convergence properties for all of them. No individual proofs are necessary, which have been always the crucial point in former investigations. Theorem 2 states the main contents of metrization theory, in particular the completeness theorem for each type of Poincaré series found in this manner. We will give two applications of our general theory; the first one is concerned with Fourier expansions, and the second one with Taylor expansions of modular forms.

In the case of Fourier expansions,

$$f(z) = \sum_{t \geq 0} a_t e^{2\pi i \sigma(tz)},$$

the relevant family of functions is

$$\{e^{2\pi i \sigma(tz)} | t \text{ half-integral}, t \geq 0\}$$

and (14) holds with the elementary formula

$$a_t = \int_{x \bmod 1} f(z) e^{-2\pi i \sigma(tz)} \, dx$$

for the Fourier coefficients. Take for Λ_n the subgroup of translations

$$\Lambda_n = \left\{ \begin{pmatrix} \pm 1 & s \\ 0 & \pm 1 \end{pmatrix} \middle| \, s \text{ symmetric with integral entries} \right\}.$$

Then according to Proposition 3 we first have to determine the Fourier expansion of

$$\sum_s \det(z + w + s)^{-k}$$

with respect to w. For this purpose we need the following generalization of Euler's Γ-integral.

Lemma 2

Let k, n be positive integers, $k > (n-1)/2$ and z any n-rowed complex symmetric matrix with positive real part. Then

$$\int_{t>0} \det t^{k-(n+1)/2} e^{-\sigma(tz)}\, dt = \pi^{n(n-1)/4} \prod_{\nu=1}^{n} \Gamma\left(k - \frac{\nu-1}{2}\right) \det z^{-k},$$

where the integral is taken over the space of all n-rowed positive definite matrices t.

Proof

We argue by induction on n. For $n = 1$ the formula

$$\int_0^\infty t^{k-1} e^{-tz}\, dt = \frac{\Gamma(k)}{z^k} \qquad (\operatorname{Re} k > 0, \operatorname{Re} z > 0) \tag{17}$$

is well known. Assume the lemma to be valid for $n-1$ and consider the n-rowed case. The assertion is invariant with respect to the replacement $z \mapsto z[a]$, where a is any non-singular real matrix. Since $\operatorname{Re} z > 0$, the matrix z can be transformed into diagonal form by an appropriate matrix a. So we may assume z to be a diagonal matrix. Decompose

$$t = \begin{pmatrix} t_1 & 0 \\ 0 & t_2 \end{pmatrix}\left[\begin{pmatrix} 1 & t_3 \\ 0 & 1 \end{pmatrix}\right], \qquad z = \begin{pmatrix} z_1 & 0 \\ 0 & z_2 \end{pmatrix},$$

where t_1 and z_1 are $(n-1)$-rowed. Because of

$$dt = \det t_1\, dt_1\, dt_2\, dt_3$$

we obtain

$$\int_{t>0} \det t^{k-(n+1)/2} e^{-\sigma(tz)}\, dt$$

$$= \int_{t_1>0, t_2>0, t_3} \det t_1^{k-(n-1)/2} e^{-\sigma(t_1 z_1)} t_2^{k-(n+1)/2} e^{-(t_2 + t_1[t_3])z_2}\, dt_1\, dt_2\, dt_3.$$

The integral over t_2 is one-dimensional and has the value

$$\int_{t_2>0} t_2^{k-(n+1)/2} e^{-t_2 z_2}\, dt_2 = \frac{\Gamma\left(k - \dfrac{n-1}{2}\right)}{z_2^{k-(n-1)/2}}$$

according to (17). The integration over t_3 yields

$$\int_{t_3} e^{-t_1[t_3]z_2}\,dt_3 = \det t_1^{-1/2}\int_h e^{-{}^t h\, h z_2}\,dh$$
$$= \det t_1^{-1/2}\left(\frac{\pi}{z_2}\right)^{(n-1)/2}.$$

Hence we obtain

$$\int_{t>0} \det t^{k-(n+1)/2} e^{-\sigma(tz)}\,dt$$
$$= \pi^{(n-1)/2} z_2^{-k}\Gamma\left(k-\frac{n-1}{2}\right)\int_{t_1>0}\det t_1^{k-n/2} e^{-\sigma(t_1 z_1)}\,dt_1,$$

and the induction hypothesis may be applied to the integral over t_1, thus proving the lemma for n.

We now apply Poisson's summation formula to the function

$$f(t) = \begin{cases}\det t^{k-(n+1)/2} e^{2\pi i\sigma(tz)} & \text{for } t>0\\ 0 & \text{otherwise}\end{cases}$$

of the real symmetric matrix-variable t. Formally we obtain

$$\sum_{t>0} \det t^{k-(n+1)/2} e^{2\pi i\sigma(tz)}$$
$$= 2^{n(n-1)/2}\sum_s \int_{t>0}\det t^{k-(n+1)/2} e^{2\pi i\sigma(t(z+s))}\,dt.$$

Here t runs over all half-integral positive definite, and s over all integral symmetric matrices. The application of Lemma 2 yields

$$\sum_{t>0} \det t^{k-(n+1)/2} e^{2\pi i\sigma(tz)} = b_{nk}\sum_s \det(z+s)^{-k}, \tag{18}$$

where

$$b_{nk} = (4\pi)^{n(n-1)/4}(-2\pi i)^{-kn}\prod_{\nu=1}^{n}\Gamma\left(k-\frac{\nu-1}{2}\right).$$

Note that on the right-hand side of (18) a subseries of the Eisenstein series $E_{n,0}^k$ appears, and on the left-hand side a Riemann sum of the integral in Lemma 2. Formula (18) turns out to be valid for $k > n+1$ in accordance with the convergence conditions for the Eisenstein series in Theorem 5.1 and for the integral in Lemma 2.

In the case of Fourier expansions Proposition 3 and Theorem 2 yield the following results. From (18) we obtain

$$\sum_s \det(z+w+s)^{-k} = b_{nk}^{-1}\sum_{t>0}\det t^{k-(n+1)/2} e^{2\pi i\sigma(t(z+w))}$$

as an explication of formula (15) in Proposition 3. Hence our general method leads to the Poincaré series

$$b_{nk}^{-1} \det t^{k-(n+1)/2} g_n^k(z,t)$$

for half-integral positive t, where

$$g_n^k(z,t) = \sum_{m \in A_n \backslash \Gamma_n} e^{2\pi i \sigma(tm\langle z\rangle)} j(m,z)^{-k}$$

is the Poincaré series of exponential type introduced by H. Maass in [49]. The main conclusions are: for $k > 2n$ and $kn \equiv 0 \bmod 2$

(i) the Poincaré series $g_n^k(z,t)$ converge absolutely and uniformly on vertical strips of positive height and represent cusp forms,

(ii) $\det y^{k/2} \left| g_n^k(z,t) \right|$ is bounded in H_n,

(iii) $P_n^k(z,w) = b_{nk}^{-1} \sum_{t>0} \det t^{k-(n+1)/2} g_n^k(z,t) e^{2\pi i \sigma(tw)}$,

(iv) the main-formula of metrization theory states

$$a_t(f) = c_{nk}^{-1} \det t^{k-(n+1)/2} \{f, g_n^k(*,t)\},$$

where $a_t(f)$ is the tth Fourier coefficient of the cusp form f and $c_{nk} = a_{nk}^{-1} \bar{b}_{nk}$ is the numerical constant

$$c_{nk} = \pi^{n(n-1)/4} (4\pi)^{n(n+1)/2-nk} \prod_{v=1}^{n} \Gamma\left(k - \frac{n+v}{2}\right),$$

(v) there are finitely many $t_1, \ldots, t_s$ such that

$$\{g_n^k(*,t_v) \mid v = 1, \ldots, s\}$$

span the space S_n^k of cusp forms.

Remark

The most important statement is certainly the completeness theorem (v). We would like to point out that for this purpose alone the lengthy computations of Lemma 2 and the subsequent investigations wouldn't have been necessary. Indeed, we only have to realize that the Fourier expansion of $\sum_s \det(z+w+s)^{-k}$ is of the form

$$\sum_t c_t e^{2\pi i \sigma(t(z+w))},$$

which is trivial because of the periodicity of that function in $w+s$. Then we can already infer from Proposition 3 and Theorem 2 that the relevant Poincaré series are of the form $g_n^k(z,t)$ and that they generate the space S_n^k of cusp forms.

The second example is concerned with Taylor expansions. It is useful to introduce as 'local coordinates' at any given point $w_0 \in H_n$

$$\zeta = l_{w_0}\langle w\rangle,$$

where l_{w_0} is Cayley's transformation (1). The range of the variable ζ is the bounded domain D_n. We take for Λ_n the subgroup $\{\pm \mathbf{1}\}$ and for

$$h_\nu(w) = \frac{\varphi_\nu(\zeta)}{\det(w - \bar{w}_0)^k},$$

where φ_ν runs over all power products in the elements of ζ. Then we understand

$$f(w) = \det(w - \bar{w}_0)^{-k} \sum_\nu a_\nu(f, w_0)\varphi_\nu(\zeta) \tag{19}$$

to be the Taylor expansion of the cusp form $f \in S_n^k$ at the given point $w_0 \in H_n$. For the integral transformation (14) we can take Cauchy's integral formula. According to Proposition 3 we have to determine the $(h_\nu^\#)$-expansion of $\det(z + w)^{-k}$ with respect to w. Now it is easily checked that

$$h_\nu^\#(w) = \overline{\varphi_\nu(l_{w_0}\langle -\bar{w}\rangle)} \det(w + w_0)^{-k}$$

and

$$\frac{\det(w + w_0)}{\det(z + w)} = \frac{\det(w_0 - \bar{w}_0)}{\det(z - \bar{w}_0)\det(\mathbf{1} - \overline{l_{w_0}\langle -\bar{w}\rangle} l_{w_0}\langle z\rangle)}.$$

Define the homogeneous polynomials $\psi_\nu(\eta)$ by the identity

$$\det(\mathbf{1} - \bar{\zeta}\eta)^{-k} = \sum_\nu \overline{\varphi_\nu(\zeta)}\psi_\nu(\eta). \tag{20}$$

Then we obtain as $(h_\nu^\#)$-expansion

$$\det(z + w)^{-k} = \sum_\nu \frac{\det(w_0 - \bar{w}_0)^k}{\det(z - \bar{w}_0)^k} \psi_\nu(l_{w_0}\langle z\rangle) h_\nu^\#(w).$$

So Theorem 2 leads to the family of Poincaré series

$$P_n^k(z, -\bar{w}_0; \psi_\nu) = \sum_{m \in \pm \mathbf{1}\backslash\Gamma_n} \frac{\psi_\nu(l_{w_0} m\langle z\rangle)}{\det(m\langle z\rangle - \bar{w}_0)^k j(m, z)^k}, \tag{21}$$

where the parameter ν varies and $w_0 \in H_n$ is fixed. This new family of Poincaré series satisfies a completeness theorem, and we have as metrization formula

$$a_\nu(f, w_0) = \det(w_0 - \bar{w}_0)^k a_{nk}\{f, P_n^k(*, -\bar{w}_0; \psi_\nu)\}, \tag{22}$$

which connects the Taylor coefficients of a cusp form f on the one hand and the scalar product with Poincaré series on the other. Of course, one can define the Poincaré series (21) for arbitrary polynomials in $n(n + 1)/2$ unknowns instead of ψ_ν.

Remarks

It is not necessary for the proof to know whether the Taylor expansions converge in all of D_n. Convergence in a fixed neighborhood of the origin is sufficient. Another interesting example of Poincaré series, which can be handled in this manner, comes from the Fourier–Jacobi expansions of modular forms (cf. [46]).

For later reference we list here the three families of Poincaré series which have been considered in detail:

type I: $\{P_n^k(*, w) \mid w \in H_n\}$,

type II: $\{P_n^k(*, w; \varphi) \mid \varphi$ an arbitrary polynomial, $w \in H_n$ fixed$\}$,

type III: $\{g_n^k(*, t) \mid t$ positive definite and half-integral$\}$.

The main results on metrization theory are valid for each type provided $k > 2n$, $kn \equiv 0 \bmod 2$. They are linked by the fact that type I is a common generating function for types II and III via Fourier, respectively Taylor expansion with respect to w.

The metrization formula (22) relates the coefficient $a_\nu(f, w_0)$ of φ_ν in the Taylor expansion (19) to the Poincaré series formed with the polynomial ψ_ν. In general, ψ_ν is different from φ_ν. The question arises of whether it is possible to replace the φ_ν by such polynomials that $\varphi_\nu = \psi_\nu$ becomes valid for all ν. Then the metrization formula would attain a more symmetric shape. The answer is affirmative; this result can be achieved by the following orthonormalization process. Let p_ν $(\nu = 0, 1, 2, \dots)$ be the m_ν-dimensional row of all power products of degree ν in the entries of ζ. Then

$$\int_{D_n} {}^t\bar{p}_\nu p_\mu \det(1 - \bar{\zeta}\zeta)^k \, dv_\zeta = 0$$

for $\nu \neq \mu$, as can be seen by substituting $\zeta \mapsto e^{is}\zeta$ with real s. For $\nu = \mu$ the Hermitian matrix

$$q_\nu = \int_{D_n} {}^t\bar{p}_\nu p_\nu \det(1 - \bar{\zeta}\zeta)^k \, dv_\zeta$$

is positive definite. Determine the matrices α_ν such that $q_\nu\{\alpha_\nu\} = 1$. Then the components of $p_\nu \alpha_\nu$ deliver a complete system of homogeneous polynomials of degree ν

$$\varphi_{\nu 1}, \varphi_{\nu 2}, \dots, \varphi_{\nu, m_\nu} \qquad (\nu = 0, 1, 2, \dots)$$

such that

$$\int_{D_n} \overline{\varphi_{\nu r}(\zeta)} \varphi_{\mu s}(\zeta) \det(1 - \bar{\zeta}\zeta)^k \, dv_\zeta = \delta_{\nu\mu}\delta_{rs}.$$

Now rewrite the Taylor expansion (19) as

$$f(w) = \det(w - \bar{w}_0)^{-k} \sum_{\nu=0}^{\infty} \sum_{\mu=1}^{m_\nu} a_{\nu\mu}(f; w_0) \varphi_{\nu\mu}(\zeta),$$

where $\zeta = l_{w_0}\langle w\rangle$ as before. Then it turns out that (20) holds with

$$\det(\mathbf{1} - \bar{\zeta}\eta)^{-k} = \int_{D_n} \det(\mathbf{1} - \bar{\zeta}\zeta)^k \, dv_\zeta \sum_{\nu=0}^{\infty} \sum_{\mu=1}^{m_\nu} \overline{\varphi_{\nu\mu}(\zeta)}\varphi_{\nu\mu}(\eta).$$

This formula can be derived either from general properties of the Bergman kernel or in an elementary way from the reproducing formula of Proposition 1, if transformed to D_n and applied to the polynomials $\varphi_{\nu\mu}$. So we may take

$$\psi_{\nu\mu} = \int_{D_n} \det(\mathbf{1} - \bar{\zeta}\zeta)^k \, dv_\zeta \, \varphi_{\nu\mu}.$$

The value of Hua's integral was linked to a_{nk} by

$$\int_{D_n} \det(\mathbf{1} - \bar{\zeta}\zeta)^k \, dv_\zeta = a_{nk}^{-1} \mathrm{i}^{-nk} 2^{-n(n-k+1)}.$$

Hence we obtain the metrization formula in the symmetric form

$$a_{\nu\mu}(f, w_0) = 2^{-n(n-2k+1)} \det v_0^k \{f, P_n^k(*, -\bar{w}_0; \varphi_{\nu\mu})\},$$

where $a_{\nu\mu}(f, w_0)$ is the Taylor coefficient corresponding to the same polynomial $\varphi_{\nu\mu}$ which appears in the Poincaré series on the right-hand side.

7 Non-cusp forms

We want to extend our results on cusp forms to arbitrary modular forms of large weight. For this purpose we combine the methods of §5 on Eisenstein series with the contents of §6 on cusp forms. The formation of Eisenstein series will turn out to be a reliable tool not only to lift cusp forms in fewer variables to modular forms in many variables, but even to lift analytical concepts and explicit representations of functions to the higher-dimensional case. By careful argumentation we can avoid any further investigations on convergence.

Besides the fixed groups $C_{n,r}$ and $B_{n,r}$ $(0 \le r \le n)$ of the standard boundary components introduced in §5 we need

$$A_{n,r} = \left\{ m \in \Gamma_n \middle| a = \begin{pmatrix} \pm 1 & 0 \\ * & * \end{pmatrix}, \quad c = 0, \quad d = \begin{pmatrix} \pm 1 & * \\ 0 & * \end{pmatrix} \right\}.$$

This is the subgroup of all elements of $C_{n,r}$ which induce translations on the boundary component. Obviously

$$B_{n,r} \subset A_{n,r} \subset C_{n,r}$$

holds for $0 \le r \le n$. Let $*$ denote the projections on the standard boundary component as in §5. It is easily checked that p^* runs over a complete set of representatives of $A_r\backslash\Gamma_r$ (respectively $\pm 1\backslash\Gamma_r$) if p runs over a complete set of representatives of $A_{n,r}\backslash C_{n,r}$ (respectively $B_{n,r}\backslash C_{n,r}$). Now we start with the Poincaré series $P_r^k(*, w; \varphi)$ of degree r and form the Eisenstein series $E_{n,r}^k$ attached to this Poincaré series. Use (5.10) and rewrite

$$P_r^k(z^*, w; \varphi) = \sum_{p \in B_{n,r}\backslash C_{n,r}} \varphi(l_{-\bar{w}} p\langle z\rangle^*) \det(p\langle z\rangle^* + w)^{-k} j(p, z)^{-k},$$

where $w \in H_r$ and $z = \begin{pmatrix} z^* & * \\ * & * \end{pmatrix}$ is any point in H_n with z^* as upper left $r \times r$ block. Insert this series into the Eisenstein series; after reordering we obtain

$$E_{n,r}^k(z; P_r^k(*, w; \varphi))$$
$$= \sum_{m \in C_{n,r}\backslash\Gamma_n} \sum_{p \in B_{n,r}\backslash C_{n,r}} \varphi(l_{-\bar{w}} pm\langle z\rangle^*) \det(pm\langle z\rangle^* + w)^{-k} j(p, m\langle z\rangle)^{-k} j(m, z)^{-k}$$
$$= \sum_{m \in B_{n,r}\backslash\Gamma_n} \varphi(l_{-\bar{w}} m\langle z\rangle^*) \det(m\langle z\rangle^* + w)^{-k} j(m, z)^{-k} =: P_{n,r}^k(z, w; \varphi).$$

Of course we write $P_{n,r}^k(z, w)$ if φ is the constant polynomial 1. Similarly we get

$$E_{n,r}^k(z; g_r^k(*, t^*))$$
$$= \sum_{m \in C_{n,r}\backslash\Gamma_n} \sum_{p \in A_{n,r}\backslash C_{n,r}} e^{2\pi i \sigma(tpm\langle z\rangle)} j(p, m\langle z\rangle)^{-k} j(m, z)^{-k}$$
$$= \sum_{m \in A_{n,r}\backslash\Gamma_n} e^{2\pi i \sigma(tm\langle z\rangle)} j(m, z)^{-k} =: g_{n,r}^k(z, t),$$

where we have set $t = \begin{pmatrix} t^* & 0 \\ 0 & 0 \end{pmatrix}$, $t^* > 0$. So each type of Poincaré series can be generalized for $0 \leq r \leq n$ in the following way,

type I: $\{P_{n,r}^k(*, w) | \, w \in H_r\}$,

type II: $\{P_{n,r}^k(*, w; \varphi) | \, \varphi$ an arbitrary polynomial, $w \in H_r$ fixed$\}$, (1)

type III: $\{g_{n,r}^k(*, t) | \, t = \begin{pmatrix} t^* & 0 \\ 0 & 0 \end{pmatrix}$, t^* positive definite and half-integral$\}$.

We formally take over the conditions $k > n + r + 1$ and $k \equiv 0 \bmod 2$ from the theory of Eisenstein series. In fact we have absolute convergence of the new Poincaré series under these conditions, as can be seen, without any further calculations, from Godement's theorem and the majorant (5.11) for Eisenstein series.

The new Poincaré series were derived from Eisenstein series attached to those Poincaré series which represent cusp forms. First we deduce from Proposition 5.8 and the completeness theorems of §6 that the Poincaré series (1) represent modular forms of degree n and weight k, if considered as functions of z, and that for fixed n, k, r each type spans $M^k_{n,r}$ provided $k > n + r + 1$ and k is even. This implies in particular a metrical characterization of the different kinds of Poincaré series according to the choice of n, r, k. Next we apply Theorem 5.2. If r varies from 0 to n, each type of Poincaré series separately generates the full linear space M^k_n of all modular forms of even weight $k > 2n$. From (5.17) we may deduce that the Φ-operator maps Poincaré series onto Poincaré series diminishing the degree by one and leaving all the parameters fixed; more precisely

$$P^k_{n,r}(*, w)|\Phi = P^k_{n-1,r}(*, w), \qquad P^k_{n,r}(*, w; \varphi)|\Phi = P^k_{n-1,r}(*, w; \varphi),$$

$$g^k_{n,r}(*, t)|\Phi = g^k_{n-1,r}(*, t_1) \qquad (0 \le r < n,\ k > n + r + 1,\ k \text{ even}),$$

where t_1 is obtained from t by deleting the last row and column. Then we consider $P^k_{n,r}(z, w)$ as a function of w. We know from the corollary of Proposition 6.2 that $P^k_r(z, w)$ represents not only a cusp form with respect to z but also a cusp form with respect to w. Since S^k_r is of finite dimension we may decompose this function into

$$P^k_r(z, w) = \sum_{\nu=1}^{l} f_\nu(z) g_\nu(w),$$

where f_ν and g_ν are cusp forms of degree r and weight k. Form the Eisenstein series with respect to z on both sides of this equation. Then we obtain for $k > n + r + 1$ and $k \equiv 0 \bmod 2$

$$P^k_{n,r}(z, w) = \sum_{\nu=1}^{l} E^k_{n,r}(z; f_\nu) g_\nu(w).$$

Hence $P^k_{n,r}(z, w)$ turns out to be a cusp form with respect to w for any fixed z.

So far we did not use anything about uniform convergence of the Poincaré series in the case $r < n$. Nevertheless they are uniformly convergent in vertical strips of positive height provided $k > n + r + 1$ and k is even. A direct proof along the lines formerly used in similar cases and working with methods from symplectic geometry is possible, but rather tedious. Hence we prefer another argument by which everything can be reduced to our former investigations. The following lemma is due to F. Berger (unpublished).

Lemma

For $0 < r < n$, $k > n + r + 1$ and $k \equiv 0 \bmod 2$ the Poincaré series $P^k_{n,r}(z, w)$ is a subseries of Siegel's Eisenstein series

$$E^k_{n+r,0}\begin{pmatrix} w & 0 \\ 0 & z \end{pmatrix} \qquad (z \in H_n,\ w \in H_r).$$

Proof

Take $m = \begin{pmatrix} a & b \\ c & d \end{pmatrix} \in \Gamma_n, \begin{pmatrix} 0 & 1 \\ -1 & 0 \end{pmatrix} \in \Gamma_r$ and form the following elements of Γ_{n+r}:

$$m_1 = \left(\begin{array}{cc|cc} 0 & 0 & -1 & 0 \\ 0 & 1 & 0 & 0 \\ \hline 1 & 0 & 0 & 0 \\ 0 & 0 & 0 & 1 \end{array}\right), \qquad m_3 = \left(\begin{array}{cc|cc} 1 & 0 & 0 & 0 \\ 0 & a & 0 & b \\ \hline 0 & 0 & 1 & 0 \\ 0 & c & 0 & d \end{array}\right),$$

$$m_2 = \left(\begin{array}{cc|c} 1 & {}^tq & \\ 0 & 1 & 0 \\ \hline & 0 & * \end{array}\right), \qquad q = \begin{pmatrix} 1 \\ 0 \end{pmatrix}.$$

Then the cocycle relation for $\hat{m} = m_1 m_2 m_3$ yields

$$j\left(\hat{m}, \begin{pmatrix} w & 0 \\ 0 & z \end{pmatrix}\right) = j\left(m_1 m_2, m_3 < \begin{pmatrix} w & 0 \\ 0 & z \end{pmatrix} > \right) j\left(m_3, \begin{pmatrix} w & 0 \\ 0 & z \end{pmatrix}\right),$$

and

$$j\left(m_1 m_2, \begin{pmatrix} w & 0 \\ 0 & z \end{pmatrix}\right) = \det\begin{pmatrix} w & {}^tqz \\ -q & 1 \end{pmatrix} = \det(w + z[q]),$$

$$j\left(m_3, \begin{pmatrix} w & 0 \\ 0 & z \end{pmatrix}\right) = \det(cz + d).$$

Therefore

$$j\left(\hat{m}, \begin{pmatrix} w & 0 \\ 0 & z \end{pmatrix}\right) = \det(m\langle z\rangle^* + w)\det(cz + d),$$

which relates the terms of $P^k_{n,r}(z, w)$ with those of $E^k_{n+r,0}\begin{pmatrix} w & 0 \\ 0 & z \end{pmatrix}$. It is a trivial matter to show that different terms of $P^k_{n,r}(z, w)$ yield different terms of $E^k_{n+r,0}\begin{pmatrix} w & 0 \\ 0 & z \end{pmatrix}$.

Corollary

For $0 \le r \le n$, $k > n + r + 1$ and $k \equiv 0 \bmod 2$ the Poincaré series $P^k_{n,r}(z, w)$ converge uniformly with respect to z, w in the product of two vertical strips of positive height, more precisely

$$\overline{|P^k_{n,r}(z, w)|} \prec \overline{|P^k_{n,r}(i1, i1)|}$$

uniformly in $z \in V_n(d)$, $w \in V_r(d)$ for any fixed $d > 0$.

The corollary follows from the corresponding fact for Siegel's Eisenstein series proved in Lemma 5.2 and the lemma above.

Since $P^k_{n,r}(*, w)$ is a cusp form of degree r with respect to w, we can form the scalar product with any other cusp form of the same degree and weight. The result is the following integral formula, which describes the inverse map $\psi_{n,r}$ of

$$\Phi^{n-r}: M^k_{n,r} \to S^k_r$$

in a modified way.

Proposition 1

Let $0 \leq r \leq n$, $k > n + r + 1$ and k be even. For any $f \in S^k_r$ the function

$$F(z) = a_{rk}\{f, P^k_{n,r}(-\bar{z}, *)\} \qquad (z \in H_n)$$

is the uniquely determined modular form in $M^k_{n,r}$, which satisfies $F|\Phi^{n-r} = f$.

Proof

By Theorem 6.1 any $f \in S^k_r$ satisfies the integral equation

$$\begin{aligned} f(z) &= a_{rk}\{f, P^k_r(-\bar{z}, *)\} \\ &= a_{rk} \int_{F_r} f(w)\, \overline{P^k_r(-\bar{z}, w)} \det(\operatorname{Im} w)^k \, dv_w. \end{aligned}$$

Forming the Eisenstein series on both sides yields

$$E^k_{n,r}(z; f) = a_{rk} \int_{F_r} f(w)\, \overline{P^k_{n,r}(-\bar{z}, w)} \det(\operatorname{Im} w)^k \, dv_w. \tag{2}$$

Here we have interchanged summation and integration, which is legitimate because of the uniform convergence of $P^k_{n,r}(-\bar{z}, w)$ with respect to w in any vertical strip of positive height. The proposition follows now from our former results on Eisenstein series proved in §5.

To shorten the notation, call the integral operator of the proposition $I_{n,r}$, and the formation of Eisenstein series $\psi_{n,r}$. Then (2) may be rewritten as

$$\psi_{n,r}(f) = I_{n,r}(f)$$

for all cusp forms f. Here we assume the weight k to be fixed. But, whereas Eisenstein series attached to non-cusp forms in general do not converge, the integral operator $I_{n,r}$ makes sense for any modular form f since $P^k_{n,r}(z, w)$ is a cusp form with respect to the second argument w. Hence we may extend the meaning of the integral operator to a mapping

$$I_{n,r}: M^k_r \to M^k_n$$

defined throughout M_r^k by the original formula. The kernel of this linear mapping is obviously N_r^k, the orthogonal complement of the space of cusp forms. Looking at the diagram of Theorem 5.2 one may then describe the inverse map of

$$\Phi\colon N_n^k \to M_{n-1}^k$$

explicitly by

$$\Phi^{-1} = \sum_{r=0}^{n-1} \Phi^{n-r-1} I_{n,r}$$

for any $k > 2n$, $k \equiv 0 \bmod 2$.

Finally we transfer the analytical connection between the different types of Poincaré series, found for cusp forms in §6, to the general situation.

Proposition 2

For $0 \le r \le n$ and even $k > n + r + 1$ the Fourier expansion of $P_{n,r}^k(z, w)$ with respect to w is

$$P_{n,r}^k(z, w) = b_{rk}^{-1} \sum_{t>0} \det t^{k-(r+1)/2} g_{n,r}^k\left(z, \begin{pmatrix} t & 0 \\ 0 & 0 \end{pmatrix}\right) e^{2\pi i \sigma(tw)},$$

where t runs over all r-rowed positive half-integral matrices.

Proof

We have proved this formula for $n = r$ in §6. Hence,

$$\int_{\operatorname{Re} w \bmod 1} P_r^k(z, w) e^{-2\pi i \sigma(tw)}\, dw = b_{rk}^{-1} \det t^{k-(r+1)/2} g_r^k(z, t).$$

Form again the Eisenstein series with respect to z on both sides and interchange summation and integration to obtain the right formula for the Fourier coefficients of $P_{n,r}^k(z, w)$ as suggested by the proposition.

A similar argument holds for Taylor expansions. So $P_{n,r}^k(z, w)$ is a common generating function of the Poincaré series of types II and III via Fourier, respectively Taylor expansion with respect to w.

Remark

One can use this connection to prove the absolute and uniform convergence of the Poincaré series $g_{n,r}^k(z, t)$ in vertical strips of positive height from the corresponding properties of $P_{n,r}^k(z, w)$.

IV

Small weights

8 Singular modular forms and theta-series

For large weights Eisenstein and Poincaré series were an appropriate method of constructing modular forms. But for small weights these series diverge and so we have to look for another method. In the one-variable case theta-functions are at least as important as Eisenstein or Poincaré series. Although theta-series in several variables appeared early in Siegel's analytic theory of quadratic forms, rather little attention was paid to them during an extended period afterwards. Only much later theta-functions did gain more respect by the impressive work of A.N. Andrianov [2], M. Eichler [17], E. Freitag [24] and J.-I. Igusa [37], amongst others, and nowadays their outstanding importance is acknowledged without any doubt. The significance of theta-series is rooted in number theory. There are no difficulties concerned with convergence or non-vanishing theorems; on the other hand the transformation formula is difficult and in general only automorphic forms for subgroups of the modular group arise. Finally, theta-functions do not fit into Petersson's metrization theory. In this section we deal with theta-series only to the extent necessary for the theory of singular forms, and we restrict ourselves to just a few remarks about recent developments at the end of the section.

Singular forms are introduced as counterparts to cusp forms by the following

Definition 1

A modular form f of degree n and weight k is called singular if its Fourier coefficients $a(t)$ vanish for all half-integral positive t.

Hence the summation in the Fourier expansion

$$f(z) = \sum_{t \geq 0} a(t) e^{2\pi i \sigma(tz)}$$

of a singular form may be restricted to singular t. The significance of singular forms is hidden in the one-variable case, since then singular forms are exactly the constant functions. On the other hand we call any non-negative integer $k < n/2$ a singular weight for modular forms of degree n.

H. Maass proved in [51] that each singular form has singular weight by using differential operators in weakly symmetric Riemannian spaces. The converse, that each modular form of singular weight is singular, was proved by H.L. Resnikoff [59] and E. Freitag [21]. We deduce both results in the first part of this section by studying the so-called Fourier–Jacobi expansions of modular forms. The structure theorem discovered by Freitag [22] and S. Raghavan [58] says that the space of singular forms is spanned by theta-series. Nowadays this can be proved independently of the characterization of singular forms by their weights. We reproduce the proof of Freitag [23] in the second part of this section.

Among the first examples of modular forms which appeared in the literature were the analytical class-invariants in Siegel's theory of quadratic forms. These invariants were defined by the following theta-series.

Definition 2

Let n, l be positive integers, t a half-integral positive $l \times l$ matrix, and $z \in H_n$. Then

$$\vartheta(z,t) = \sum_g e^{2\pi i\sigma(tz[g])},$$

where g runs over all $n \times l$ matrices with integral entries.

Since t and $y = \operatorname{Im} z$ are positive, theta-series converge very well and represent holomorphic functions in z on H_n. Obviously $\vartheta(*, t)$ is invariant with respect to integral modular substitutions. The behavior with respect to a bigger subgroup of the modular group can be derived from the inversion formula for theta-series and is stated here without proof.

Proposition 1

$\vartheta(z,t)$ is a modular form with respect to

$$\Gamma_0(q) = \{m \in \Gamma_n | c \equiv 0 \bmod q\}$$

of weight $l/2$ and with certain eighth roots of unity as multiplier system. Here q is the level of the quadratic form $2t$, i.e. the smallest positive integer such that $q(2t)^{-1}$ becomes integral. In particular $\vartheta(z,t)$ is a modular form if and only if $q = 1$.

We have assumed t to be half-integral. In the theory of quadratic forms it is more common to say equivalently $s = 2t$ is even, since only even numbers are represented by such a form. Looking for modular forms among the theta-series means searching for positive even quadratic forms of determinant one. The existence of such forms is by no means trivial (cf.

J.-P. Serre [61]). H. Minkowski found the first example for $l = 8$. It is known that such forms exist if and only if $l \equiv 0 \bmod 8$. For $l = 8$ we have, for instance,

$$2x_1^2 + 2x_2^2 + 4x_3^2 + 4x_4^2 + 20x_5^2 + 12x_6^2 + 4x_7^2 + 2x_8^2 +$$
$$+ 2x_1x_2 + 2x_2x_3 + 6x_3x_4 + 10x_4x_5 + 6x_5x_6 + 2x_6x_7 + 2x_7x_8,$$

and taking direct sums yields examples for any $l \equiv 0 \bmod 8$. From the definition we immediately deduce the Fourier expansion

$$\vartheta(z, t) = \sum_{s \geq 0} \alpha(s, t) e^{2\pi i \sigma(sz)},$$

where

$$\alpha(s, t) = \#\{g \mid s = t[g], g \text{ integral}\}$$

denotes the number of representations of the quadratic form s by t. Now assume $l < n$. Since $\operatorname{rank} s \leq \operatorname{rank} t$ if s is representable by t, we obtain $\alpha(s, t) = 0$ for any non-singular s. Hence $\vartheta(z, t)$ represents a singular form of weight $l/2$ and degree n for any even quadratic form $2t$ of determinant one and order $l < n$. So the class-invariants of C.L. Siegel yield non-trivial examples of singular forms of any weight $k \equiv 0 \bmod 4$, $0 < k < n/2$.

First we want to characterize singular forms by their weights, which is rather hard. The reader who is mainly interested in the structure theorem for singular forms may omit the following investigations and pass on to the end of this section. We consider the Fourier–Jacobi expansions of modular forms which were introduced by I.-I. Pjateckij-Šapiro [56] in the following manner.

Definition 3

Let $0 < r < n$ and decompose the $n \times n$ matrices z, t into

$$z = \begin{pmatrix} z_1 & z_2 \\ * & z_4 \end{pmatrix}, \qquad t = \begin{pmatrix} t_1 & t_2 \\ * & t_4 \end{pmatrix},$$

where z_1, t_1 have r rows and columns. Then for any modular form $f \in M_n^k$ the reordering of its Fourier expansion

$$f(z) = \sum_{t_4 \geq 0} \left(\sum_{t_1, t_2; t \geq 0} a(t) e^{2\pi i \sigma(t_1 z_1 + 2t_2 {}^t z_2)} \right) e^{2\pi i \sigma(t_4 z_4)}$$
$$= \sum_{t_4 \geq 0} \beta(z_1, z_2; t_4) e^{2\pi i \sigma(t_4 z_4)} \tag{1}$$

is called the Fourier–Jacobi expansion of type $(r, n - r)$.

The Fourier–Jacobi coefficients $\beta(z_1, z_2; t_4)$ represent holomorphic functions on $H_r \times \mathbb{C}^{r(n-r)}$, since any pair (z_1, z_2), $z_1 \in H_r$, can be complemented to an element z of H_n. Here we have identified the space of $r \times (n-r)$ matrices z_2 with the complex number space $\mathbb{C}^{r(n-r)}$. The transformation formula for modular forms implies a certain behavior of the Fourier–Jacobi coefficients which we want to derive. For given $m_1 = \begin{pmatrix} a & b \\ c & d \end{pmatrix} \in \Gamma_r$ complete the blocks a, b, c, d to the $n \times n$ matrices

$$\begin{pmatrix} a & 0 \\ 0 & 1 \end{pmatrix}, \quad \begin{pmatrix} b & 0 \\ 0 & 0 \end{pmatrix}, \quad \begin{pmatrix} c & 0 \\ 0 & 0 \end{pmatrix}, \quad \begin{pmatrix} d & 0 \\ 0 & 1 \end{pmatrix}$$

and form with these new blocks an element $m \in \Gamma_n$, denoted by $m = m_1 * 1$ for short. Then

$$m\langle z \rangle = \begin{pmatrix} m_1\langle z_1 \rangle & {}^t(cz_1 + d)^{-1} z_2 \\ * & z_4 - {}^t z_2 (cz_1 + d)^{-1} c z_2 \end{pmatrix},$$

and the transformation formula of f with respect to m yields

$$\begin{aligned} &\beta(m_1\langle z_1 \rangle, {}^t(cz_1 + d)^{-1} z_2; t_4) \\ &\quad = \det(cz_1 + d)^k e^{2\pi i \sigma(t_4 {}^t z_2 (cz_1 + d)^{-1} c z_2)} \beta(z_1, z_2; t_4) \end{aligned} \tag{2}$$

for any $m_1 \in \Gamma_r$. Hence the behavior of β as a function of z_1 does not differ too much from that of a modular form of degree r and weight k. Then obviously β is periodic with respect to z_2 of period one,

$$\beta(z_1, z_2 + g; t_4) = \beta(z_1, z_2; t_4) \tag{3}$$

for all integral g. Finally apply the modular substitution

$$z \mapsto z\left[\begin{pmatrix} 1 & g \\ 0 & 1 \end{pmatrix}\right]$$

to f in order to obtain

$$\beta(z_1, z_2 + z_1 g; t_4) = e^{-2\pi i \sigma\{t_4(z_1[g] + 2\,{}^t g z_2)\}} \beta(z_1, z_2; t_4), \tag{4}$$

again for arbitrary integral g. Formulas (3) and (4) show that β, when considered as a function of z_2, behaves like a Jacobian function in the theory of Abelian functions. Hence we are justified in calling any holomorphic function on $H_r \times \mathbb{C}^{r(n-r)}$ satisfying (2)–(4) a Jacobi form of weight k and index t_4.

Consider first only (3) and (4) and denote the corresponding class of functions by

$$A(t) = \{h(z_1, z_2) \mid h \text{ holomorphic on } H_r \times \mathbb{C}^{r(n-r)};\ h \text{ satisfies (3), (4)}\}$$

($0 \le r \le n$, t any fixed $(n-r)$-rowed positive half-integral matrix). Obviously $A(t)$ has the structure of a module over the ring of holomorphic functions in z_1 ranging over H_r. We want to determine a basis of this module. One may generalize the theta-series of Definition 2 by

$$\Theta(z_1, z_2; t, q) = \sum_g e^{2\pi i\sigma\{t(z_1[g+q]+2\,{}^t(g+q)z_2)\}}, \tag{5}$$

where the assumptions on z_1, z_2 and t are as above, and q is a complex matrix of type $(r, n-r)$. Then

$$\Theta(z_1, z_2; t, \tfrac{1}{2}at^{-1}) \in A(t) \tag{6}$$

for any integral matrix a of r rows and $n-r$ columns. Indeed, (3) is obviously satisfied and the verification of (4) is straightforward:

$$\begin{aligned}&\Theta(z_1, z_2 + z_1 h; t, \tfrac{1}{2}at^{-1})\\ &\quad= \sum_g e^{2\pi i\sigma\{t(z_1[g+(1/2)at^{-1}]+2\,{}^t(g+(1/2)at^{-1})(z_2+z_1h))\}}\\ &\quad= e^{-2\pi i\sigma\{t(z_1[h]+2\,{}^thz_2)\}}\sum_g e^{2\pi i\sigma\{t(z_1[g+h+(1/2)at^{-1}]+2\,{}^t(g+h+(1/2)at^{-1})z_2)\}}\\ &\quad= e^{-2\pi i\sigma\{t(z_1[h]+2\,{}^thz_2)\}}\Theta(z_1, z_2; t, \tfrac{1}{2}at^{-1})\end{aligned}$$

for any integral h. The theta-series in (6) do not change if a is replaced by $a + 2gt$ for any integral g. Call two integral matrices a and a^* of type $(r, n-r)$ congruent modulo $2t$ if there exists an integral g such that

$$a^* = a + 2gt.$$

The number of congruence classes is $\det(2t)^r$. We are now able to state

Proposition 2

Let $0 < r < n$ and let t be any fixed half-integral positive matrix of $n-r$ rows. Then the $\det(2t)^r$ *theta-series*

$$\{\Theta(z_1, z_2; t, \tfrac{1}{2}at^{-1}) \mid a \bmod 2t\} \tag{7}$$

form a basis of the module $A(t)$ over the ring of holomorphic functions in $z_1 \in H_r$.

Proof

The theta-series $\Theta(z_1, z_2; t, \frac{1}{2}at^{-1})$ belong to $A(t)$ by (6). We realize from their Fourier expansions with respect to z_2 that they are linear independent over the ring of holomorphic functions in z_1. Now let $h(z_1, z_2)$ be any element of $A(t)$. Since h is periodic with respect to z_2 of period one we have a Fourier expansion

$$h(z_1, z_2) = \sum_a c(z_1, a)e^{2\pi i\sigma({}^taz_2)},$$

where the matrix a runs over all integral matrices of type $(r, n-r)$. The transformation formula (4) yields for the Fourier coefficients

$$c(z_1, a)e^{2\pi i\sigma({}^taz_1g)} = c(z_1, a + 2gt)e^{-2\pi i\sigma\{t(z_1[g])\}},$$

g being any integral matrix of type $(r, n-r)$. Hence,

$$\begin{aligned} h(z_1, z_2) &= \sum_{a \bmod 2t} c(z_1, a) \sum_g e^{2\pi i \sigma\{t(z_1[g]) + {}^t a z_1 g + {}^t(a+2gt) z_2\}} \\ &= \sum_{a \bmod 2t} c(z_1, a) e^{-(\pi i/2)\sigma\{t^{-1}(z_1[a])\}} \\ &\quad \times \sum_g e^{2\pi i \sigma\{t(z_1[g+(1/2)at^{-1}] + 2\,{}^t(g+(1/2)at^{-1}) z_2)\}} \\ &= \sum_{a \bmod 2t} c(z_1, a) e^{-(\pi i/2)\sigma\{t^{-1}(z_1[a])\}} \Theta(z_1, z_2; t, \tfrac{1}{2}at^{-1}). \end{aligned}$$

Using the basis (7) of $A(t)$ we may rewrite the main part of the Fourier–Jacobi expansion (1) in the following manner. Join together the basis elements (7) to the column vector

$$\vec{\Theta}(z_1, z_2; t)$$

and put

$$\psi(z, t) = \vec{\Theta}(z_1, z_2; t) e^{2\pi i \sigma(t z_4)}$$

for any $z \in H_n$. Then, provided $t_4 > 0$, the Fourier–Jacobi coefficients of Definition 3 have the form

$$\beta(z_1, z_2; t_4) = {}^t p(z_1, t_4) \vec{\Theta}(z_1, z_2; t_4), \tag{8}$$

where $p(z_1, t_4)$ is a column vector of holomorphic functions in z_1, and the corresponding part of the Fourier–Jacobi expansion (1) is

$$\sum_{t_4 > 0} \beta(z_1, z_2; t_4) e^{2\pi i \sigma(t_4 z_4)} = \sum_{t_4 > 0} {}^t p(z_1, t_4) \psi(z, t_4).$$

Note that we have to assume $t_4 > 0$, since otherwise the theta-series may diverge. To obtain more information about this part of the Fourier–Jacobi expansion we investigate $\psi(z, t_4)$.

Proposition 3

Let $0 < r < n$, t a half-integral positive matrix of $n - r$ rows and columns and $z \in H_n$. Then the components of $\psi(z, t)$ are modular forms with respect to the group

$$\Gamma_{n,r}(q) = \{m_1 * 1 \in \Gamma_n \mid m_1 \in \Gamma_r, a_1 \equiv d_1 \equiv 1 \bmod q, b_1 \equiv 0 \bmod q^2, c_1 \equiv 0 \bmod q\}$$

of weight $(n-r)/2$ and with certain eighth roots of unity as multipliers. Here q denotes the level of the even quadratic form $2t$.

Proof

The components of $\psi(z, t)$ are

$$\Theta(z_1, z_2; t, \tfrac{1}{2}at^{-1}) e^{2\pi i \sigma(t z_4)} \tag{9}$$

with integral matrices a of a type $(r, n-r)$. On the other hand consider the theta-series

$$\vartheta(z[{}^t v], t), \qquad v = \begin{pmatrix} 1 & 0 \\ (2t)^{-1}\,{}^t a & 1 \end{pmatrix} \tag{10}$$

formed with the same matrices a. The component (9) of $\psi(z, t)$ will turn out to occur among the Fourier–Jacobi coefficients of $\vartheta(z[{}^t v], t)$; then we can apply the transformation formula of Proposition 1 to (10) and get from there the desired information about the component (9). This is the idea of the proof.

It is easily checked that

$$z[{}^t v] = \begin{pmatrix} z_1 & z_2 + z_1 a(2t)^{-1} \\ * & z_4 + z_1[a(2t)^{-1}] + {}^t z_2 a(2t)^{-1} + (2t)^{-1}\,{}^t a z_2 \end{pmatrix},$$

hence

$$\vartheta(z[{}^t v], t) = \sum_{g_1, g_2} e^{2\pi i \sigma(tp)},$$

where

$$p = z_1[g_1] + 2\,{}^t g_1(z_2 + z_1 a(2t)^{-1})g_2 + (z_4 + {}^t z_2 a t^{-1} + z_1[a(2t)^{-1}])[g_2]$$

and g_1, g_2 run over all integral matrices of type $(r, n-r)$, respectively $(n-r, n-r)$. Reorder this series as the Fourier–Jacobi expansion

$$\vartheta(z[{}^t v], t) = \sum_{s \geq 0} \alpha(z_1, z_2; s) e^{2\pi i \sigma(s z_4)}. \tag{11}$$

Then we obtain in particular for $s = t$

$$\begin{aligned} \alpha(z_1, z_2; t) &= e^{2\pi i \sigma({}^t a z_2 + (1/4) z_1[a] t^{-1})} \sum_{g_1, g_2; t[{}^t g_2] = t} e^{2\pi i \sigma\{t(z_1[g_1] + 2\,{}^t g_1(z_2 + z_1 a(2t)^{-1})g_2)\}} \\ &= e^{2\pi i \sigma({}^t a z_2 + (1/4) z_1[a] t^{-1})} \varepsilon(t) \sum_{g} e^{2\pi i \sigma\{t(z_1[g] + 2\,{}^t g(z_2 + z_1 a(2t)^{-1}))\}} \\ &= \varepsilon(t)\Theta(z_1, z_2; t, \tfrac{1}{2} a t^{-1}). \end{aligned} \tag{12}$$

Here $\varepsilon(t)$ denotes the number of integral solutions g of $t[g] = t$. Consider now any $m = m_1 * \mathbf{1} \in \Gamma_{n,r}(q)$. We have

$$\vartheta(m\langle z\rangle[{}^t v], t) = \vartheta(\tilde{m}\langle z[{}^t v]\rangle, t),$$

where

$$\tilde{m} = \begin{pmatrix} v & 0 \\ 0 & {}^t v^{-1} \end{pmatrix} m \begin{pmatrix} v^{-1} & 0 \\ 0 & {}^t v \end{pmatrix}.$$

Now $m \in \Gamma_{n,r}(q)$ implies $\tilde{m} \in \Gamma_0(q)$, as can be verified from

$$\tilde{m} = \left(\begin{array}{cc|cc} a_1 & 0 & b_1 & b_1 a(2t)^{-1} \\ (2t)^{-1}\,{}^t a(a_1 - 1) & \mathbf{1} & (2t)^{-1}\,{}^t a b_1 & b_1[a(2t)^{-1}] \\ \hline c_1 & 0 & d_1 & (d_1 - 1)a(2t)^{-1} \\ 0 & 0 & 0 & \mathbf{1} \end{array}\right).$$

So we may apply Proposition 1 and obtain

$$\vartheta(m\langle z\rangle[{}^t\mathfrak{v}],t) = \mathrm{v}(m)\det(c_1 z_1 + d_1)^{(n-r)/2}\vartheta(z[{}^t\mathfrak{v}],t)$$

with certain eighth roots of unity as multipliers $\mathrm{v}(m)$. Insert the Fourier–Jacobi expansion (11) for $\vartheta(z[{}^t\mathfrak{v}],t)$ and compare the tth Fourier–Jacobi coefficient on both sides. Using (12) we get

$$\Theta(m\langle z\rangle_1, m\langle z\rangle_2; t, \tfrac{1}{2}at^{-1})e^{2\pi i\sigma(tm\langle z\rangle_4)}$$
$$= \mathrm{v}(m)\det(cz+d)^{(n-r)/2}\Theta(z_1,z_2;t,\tfrac{1}{2}at^{-1})e^{2\pi i\sigma(tz_4)}.$$

Remark
We do not claim $\mathrm{v}(m)$ to be a character of the underlying group. But of course this becomes true if $n-r$ is even.

The whole point of Fourier–Jacobi expansions is the following consequence.

Proposition 4
Let f be any modular form of weight k and degree n. Write its Fourier–Jacobi coefficients for positive t_4 according to (8) *as*

$$\beta(z_1,z_2;t_4) = {}^t p(z_1,t_4)\vec{\Theta}(z_1,z_2;t_4).$$

Then the components of $p(z_1,t_4)$ behave like modular forms (with certain eighth roots of unity as multipliers) of the diminished weight $k-(n-r)/2$ with respect to the subgroup

$$\{m_1\in\Gamma_r \mid a_1\equiv d_1\equiv 1 \bmod q, b_1\equiv 0 \bmod q^2, c_1\equiv 0 \bmod q\}$$

of Γ_r. Here q denotes again the level of the even quadratic form $2t_4$.

Proof
The terms

$$\beta(z_1,z_2;t_4)e^{2\pi i\sigma(t_4z_4)}$$

of the Fourier–Jacobi expansion of f transform as a modular form of weight k (cf. (2)), and the components of

$$\vec{\Theta}(z_1,z_2;t_4)e^{2\pi i\sigma(t_4z_4)}$$

transform as a modular form of weight $(n-r)/2$ (cf. Proposition 3), both with respect to $\Gamma_{n,r}(q)$. Since the components of $p(z_1,t_4)$ are determined uniquely as holomorphic functions in z_1 (cf. Proposition 2) we obtain the desired result about $p(z_1,t_4)$.

Now we turn to the characterization of singular forms by their weights.

Theorem 1

Any modular form $f \in M_n^k$ of singular weight (i.e. $k < n/2$) is singular.

Proof

We may assume $k > 0$ and $n > 2$, since otherwise f is constant. Take any modular form f of weight $k < n/2$ and divide its Fourier expansion into two parts

$$f(z) = \sum_{t \geq 0,\, t \text{ singular}} a(t)e^{2\pi i\sigma(tz)} + \sum_{t>0} a(t)e^{2\pi i\sigma(tz)}. \tag{13}$$

We have to prove that the second part vanishes. Determine $0 < r < n$ such that $k \leq (n-r)/2$ and $n - r$ is even. Rearrange (13) to become the Fourier–Jacobi expansion of type $(r, n-r)$

$$f(z) = \sum_{t_4 \geq 0} \beta(z_1, z_2; t_4)e^{2\pi i\sigma(t_4 z_4)}.$$

Then as in (1) we have

$$\beta(z_1, z_2; t_4)e^{2\pi i\sigma(t_4 z_4)} = \sum_{t_2, t_2; t = \begin{pmatrix} t_1 & t_2 \\ * & t_4 \end{pmatrix} \geq 0} a(t)e^{2\pi i\sigma(tz)}.$$

If t_4 is singular, only singular t occur on the right-hand side and we don't get a contribution to the second part of the Fourier expansion (13). If t_4 is non-singular, we have by Proposition 4

$$\beta(z_1, z_2; t_4)e^{2\pi i\sigma(t_4 z_4)} = {}^t p(z_1, t_4)\vec{\Theta}(z_1, z_2; t_4)e^{2\pi i\sigma(t_4 z_4)}$$

and the components of $p(z_1, t_4)$ are modular forms of the non-positive weight $k - (n-r)/2$. The multiplier system is a character since $n - r$ is even. Hence $p(z_1, t_4)$ vanishes if $k - (n-r)/2$ is negative and constant if $k = (n-r)/2$. But even in the latter case we don't get a contribution to the second part of (13) owing to the following observation. The components of $\vec{\Theta}(z_1, z_2; t_4)e^{2\pi i\sigma(t_4 z_4)}$ are

$$\Theta(z_1, z_2; t_4, q)e^{2\pi i\sigma(t_4 z_4)}, \qquad q = \tfrac{1}{2}at_4^{-1},$$

and the general term of this theta-series (cf. (5)) may be rewritten as

$$e^{2\pi i\sigma\{t_4(z_1[g+q] + 2\,{}^t(g+q)z_2 + z_4)\}} = e^{2\pi i\sigma(Az)},$$

where

$$A = \begin{pmatrix} 0 & 0 \\ 0 & t_4 \end{pmatrix}\left[\begin{pmatrix} 1 & g+q \\ {}^t(g+q) & 1 \end{pmatrix}\right]$$

is indeed singular.

Remark
We have used the fact that modular forms for congruence subgroups vanish for negative weights and are constant for weight zero. This is a slight generalization of our former results for the full modular group.

Theorem 2
Any singular form $f \in M_n^k$ has singular weight.

Proof
Assume f to be non-constant and determine $0 < r < n$ such that the Fourier coefficients $a(s)$ of f vanish if rank $s > n - r$ and $a(t) \neq 0$ for at least one t of rank $n - r$. Fix one of those t; after unimodular transformation we may assume

$$t = \begin{pmatrix} 0 & 0 \\ 0 & t_4 \end{pmatrix}$$

with positive t_4. Consider the Fourier–Jacobi expansion (1) of type $(r, n-r)$ and in particular its coefficient

$$\beta(z_1, z_2; t_4) = {}^t p(z_1, t_4)\vec{\Theta}(z_1, z_2; t_4).$$

Then the components $p_a(z_1, t_4)$ of $p(z_1, t_4)$ are modular forms of weight $k - (n-r)/2$ with respect to a congruence subgroup by Proposition 4 and, on the other hand, by the proof of Proposition 2,

$$p_a(z_1, t_4) = e^{-(\pi i/2)\sigma\{t_4^{-1}(z_1[a])\}} \sum_{s_1; s \geq 0} a(s) e^{2\pi i \sigma(s_1 z_1)},$$

where

$$s = \begin{pmatrix} s_1 & a/2 \\ {}^t a/2 & t_4 \end{pmatrix}.$$

Since $a(s) = 0$ for rank $s > n - r$, the sum on the right consists at most of one term, namely the term with $s_1 = t_4^{-1}[{}^t a/2]$. Hence $p_a(z_1, t_4)$ is constant, and in particular

$$p_0(z_1, t_4) = a\begin{pmatrix} 0 & 0 \\ 0 & t_4 \end{pmatrix} \neq 0.$$

A constant modular form different from zero has weight zero, hence

$$k = \frac{n-r}{2} < \frac{n}{2}.$$

The structure theorem, Theorem 3, for singular forms is independent of the preceding investigations of the weights. On the contrary we incidentally prove Theorem 2 once again.

Theorem 3

Each singular form of positive weight k and degree n is a linear combination of the class invariants $\vartheta(z,t)$; here $2t$ ranges over all even positive quadratic forms of order $2k$ and determinant one. The weight satisfies $k < n/2$.

Proof

To begin with, note that each singular modular form $f \neq 0$ has even weight. Indeed, for odd weight $f|\Phi = 0$ and the only singular cusp form is obviously $f = 0$. We need this fact subsequently to have available the invariance of f with respect to all integral modular substitutions.

(i) If $f \neq 0$ is singular of weight $k > 0$ and degree n, then $k < n/2$ and there exists a positive half-integral matrix s of $2k$ rows and $\det(2s) = 1$ such that the Fourier coefficient

$$a\begin{pmatrix} 0 & 0 \\ 0 & s \end{pmatrix} \neq 0.$$

To prove this statement let ρ be the maximal rank of those t for which $a(t) \neq 0$. Then $0 < \rho < n$; any t of rank ρ with $a(t) \neq 0$ can be represented as

$$t = \begin{pmatrix} 0 & 0 \\ 0 & s \end{pmatrix}[u],$$

where u is unimodular, and s is positive and of ρ rows. Choose t and u such that $\det s$ becomes minimal under these conditions and fix s from now on. Then we certainly have satisfied

$$a\begin{pmatrix} 0 & 0 \\ 0 & s \end{pmatrix} \neq 0, \qquad s \text{ half-integral}, \qquad s > 0.$$

Now consider the restriction of f onto $H_{n-\rho} \times H_\rho$,

$$f\begin{pmatrix} w & 0 \\ 0 & z \end{pmatrix} = \sum_{t \geq 0} \alpha_t(w) e^{2\pi i \sigma(tz)} \qquad (z \in H_\rho, w \in H_{n-\rho}),$$

where

$$\alpha_t(w) = \sum_{t_2, t_2} a\begin{pmatrix} t_1 & t_2 \\ * & t \end{pmatrix} e^{2\pi i \sigma(t_1 w)}$$

is a modular form of weight k and degree $n - \rho$. We turn our attention in particular to $\alpha_s(w)$. If any coefficient

$$a\begin{pmatrix} t_1 & t_2 \\ * & s \end{pmatrix} \neq 0$$

then $\begin{pmatrix} t_1 & t_2 \\ * & s \end{pmatrix}$ is of rank ρ because of the maximal condition for the rank. Hence,

$$\begin{pmatrix} t_1 & t_2 \\ * & s \end{pmatrix} = \begin{pmatrix} 0 & 0 \\ 0 & v \end{pmatrix}[u], \qquad u \text{ unimodular},$$

and we may infer that $\det v \leq \det s$. On the other hand $\det v = \det s$ because of the minimal condition for $\det s$. Hence v and s are equivalent with respect to the unimodular group, and we obtain

$$\alpha_s(w) = a\begin{pmatrix} 0 & 0 \\ 0 & s \end{pmatrix} \sum_t e^{2\pi i \sigma(t_1 w)},$$

where $t = \begin{pmatrix} t_1 & t_2 \\ * & 0 \end{pmatrix}$ runs over all half-integral matrices ≥ 0, which may be transformed into $\begin{pmatrix} 0 & 0 \\ 0 & s \end{pmatrix}$ by unimodular matrices. It is immediately checked that this condition is equivalent to

$$t = \begin{pmatrix} 0 & 0 \\ 0 & s \end{pmatrix}\left[\begin{pmatrix} 1 & 0 \\ g & 1 \end{pmatrix}\right],$$

where g runs over all integral matrices of type $(\rho, n - \rho)$. Hence,

$$\alpha_s(w) = a\begin{pmatrix} 0 & 0 \\ 0 & s \end{pmatrix} \vartheta(w, s).$$

Comparing the weights we infer that

$$k = \frac{\rho}{2} < \frac{n}{2}.$$

Moreover, since $\alpha_s(w)$ is a modular form, we obtain $\det(2s) = 1$ by Proposition 1.

(ii) Let $f \neq 0$ be any singular form of weight $k > 0$ and degree n. Then $k < n/2$ by part (i) of this proof. Take a complete set $s_1, \ldots, s_l$ of representatives of the classes of all positive half-integral $2k$-rowed matrices s with $\det(2s) = 1$. By classes we mean equivalence classes with respect to the action of the unimodular group. It follows from Minkowski's reduction theory that the number of these classes is finite. Set

$$f^*(z) = f(z) - \sum_{v=1}^{l} c_v \vartheta(z, s_v).$$

For the Fourier coefficients we obtain

$$a^*\begin{pmatrix} 0 & 0 \\ 0 & s_v \end{pmatrix} = a\begin{pmatrix} 0 & 0 \\ 0 & s_v \end{pmatrix} - c_v \alpha(s_v, s_v) \qquad (v = 1, \ldots, l),$$

where $\alpha(s_\nu, s_\nu)$ is the number of representations of s_ν by itself, i.e. the number of units of s_ν. Now the coefficients c_ν can be determined such that

$$a^*\begin{pmatrix}0 & 0\\ 0 & s_\nu\end{pmatrix} = 0 \qquad (\nu = 1, \ldots, l).$$

Then from statement (i) applied to the singular form f^* we infer $f^* = 0$, hence

$$f(z) = \sum_{\nu=1}^{l} c_\nu \vartheta(z, s_\nu).$$

As we have mentioned before, positive even quadratic forms $2t$ of determinant one exist if and only if their order $2k \equiv 0 \bmod 8$. So we obtain

Corollary (i)
$M_n^k = 0$ if $0 < k < n/2$ and $k \not\equiv 0 \bmod 4$.

Any cusp form which is singular must be zero. Therefore Siegel's Φ-operator defines an injective map

$$\Phi\colon M_n^k \to M_{n-1}^k$$

if M_n^k consists of singular forms. Concerning the surjectivity we first mention

$$\vartheta_n(*, t)|\Phi = \vartheta_{n-1}(*, t)$$

for any half-integral positive t with $\det(2t) = 1$; here, for clarity, we have given the symbol ϑ an index n. This is obvious from the Fourier expansion of $\vartheta(z, t)$, since the number of representations of $\begin{pmatrix}s & 0\\ 0 & 0\end{pmatrix}$ by t is equal to the number of representations of s by t. Now the structure theorem states that the spaces of singular forms are spanned by those theta-series; so we obtain

Corollary (ii)
Siegel's Φ-operator is an isomorphism if

$$k < \frac{n-1}{2}.$$

The question of whether modular forms can be represented by theta-series is very important. In this section we have given an affirmative answer for weights $k < n/2$. For general k progress has been achieved only recently. In this connection we draw the reader's interest to the theory of stable modular forms by E. Freitag [22] and to the solution of the so-called basis problem for theta-functions by S. Böcherer [8, 9] and R. Weissauer [69]. They proved that each modular form of weight $k > 2n$, $k \equiv 0 \bmod 4$, can be represented as a linear combination of theta-series. For weights $k > 2n$,

$k \not\equiv 0 \bmod 4$, theta-series with harmonic coefficients have to be taken into account in order to represent at least all cusp forms. So the importance of theta-series is by no means restricted to small weights.

9 The graded ring of modular forms of degree two

Up to now we have been mainly interested in the linear spaces M_n^k of modular forms of fixed weight k and degree n. Our main theorems were concerned with estimates for the dimension of M_n^k and with different sets of generators. There was only one result which covered the totality of all modular forms, namely the algebraic dependence of $n(n+1)/2+2$ modular forms of arbitrary weights. In order to get rid of the weights we form the direct sum of the additive groups M_n^k, $k \in \mathbb{Z}$, and consider

$$M_n = \sum_{k \in \mathbb{Z}} \oplus M_n^k$$

as a $\mathbb{Z}$-graded ring. Hence the elements of M_n are formal finite sums

$$\sum_{k \in \mathbb{Z}} f_k,$$

where $f_k \in M_n^k$ and the ring multiplication maps $M_n^k \times M_n^l$ into M_n^{k+l} by

$$(f_k, f_l) \mapsto f_k f_l.$$

A famous theorem of W.L. Baily [4] states that M_n as a $\mathbb{C}$-algebra is finitely generated or equivalently, there exist finitely many modular forms $f_1, \ldots, f_N$ such that any modular form, no matter what the weight may be, can be represented as a polynomial in $f_1, \ldots, f_N$. We cannot show this theorem here, since all the proofs in the literature (cf. [26]) are based on different compactification theories. Much more is known about the structure of this ring for $n \leq 3$. By a classical result the graded ring of modular forms in one variable is generated by the elliptic invariants g_2 and g_3, which are essentially the Eisenstein series of weights 4 and 6. J.-I. Igusa proved in [34] that the graded ring of modular forms of even weights and degree two is generated by the four Eisenstein series of weights 4, 6, 10 and 12 and that these generators are algebraically independent. Later [35, 36] he was able to extend his results to odd weights, adding a cusp form of weight 35 to the set of generators and studying the algebraic relations. For $n = 3$ we refer to S. Tsuyumine [67]. In this book we only show Igusa's first theorem by an elementary proof due to E. Freitag [20].

It will be of specific importance to concentrate on modular forms of lowest possible weight. Thus the current section fits into the topic of this chapter.

Henceforth we assume $n = 2$ and introduce the following new kind of theta-series.

Definition 1

Let $z \in H_2$ and $a, b \in \mathbb{Z}^2$ be 2×1 matrices with integral entries. Then by a theta-series of characteristic (a, b) we understand

$$\vartheta(z; a, b) := \sum_{g \in \mathbb{Z}^2} e^{\pi i (z[g + (a/2)] + {}^t b g)},$$

where g runs over all integral columns of two elements.

Remark

Of course one could subordinate the different types of theta-series under a general scheme; but this probably wouldn't improve the lucidity in our rather elementary situation.

The substitution $g \mapsto -g - a$ shows

$$\vartheta(z; a, b) = (-1)^{{}^t ab} \vartheta(z; a, b).$$

Hence $\vartheta(z; a, b) = 0$ if ${}^t ab \not\equiv 0 \bmod 2$. Furthermore we immediately verify

$$\begin{aligned} \vartheta(z; a + 2g, b) &= (-1)^{{}^t bg} \vartheta(z; a, b), \\ \vartheta(z; a, b + 2g) &= \vartheta(z; a, b) \end{aligned} \tag{1}$$

for any $g \in \mathbb{Z}^2$. Hence, to cover all non-trivial cases, we may assume the characteristic (a, b) to be reduced modulo 2 and ${}^t ab \equiv 0 \bmod 2$. Then there remain ten theta-series, which have the following characteristics

a	0 0	0 1	1 0	1 1	0 0	0 1	0 0	1 0	0 0	1 1
b	0 0	0 0	0 0	0 0	1 0	1 0	0 1	0 1	1 1	1 1

(2)

We denote by

$$\Theta(z) = \prod_{a,b} \vartheta(z; a, b) \tag{3}$$

the product of these ten theta-functions. Clearly $\Theta(z)$ is a holomorphic function on Siegel's half-space of degree two. We want to study its behavior with respect to the action of the modular group.

For this purpose we may restrict ourselves to a set of generators of Γ_2, for instance (cf. Proposition 3.6) to

$$\begin{pmatrix} 0 & 1 \\ -1 & 0 \end{pmatrix}, \quad \begin{pmatrix} 1 & s \\ 0 & 1 \end{pmatrix},$$

where s is symmetric and integral. We obtain for the translations

$$\begin{aligned}\vartheta(z+s;a,b) &= \sum_g e^{\pi i(z[g+(a/2)]+s[g+(a/2)]+{}^tbg)}\\ &= e^{(\pi i/4)s[a]}\sum_g e^{\pi i(z[g+(a/2)]+{}^t(b+sa)g+s_1g_1+s_4g_2)}\\ &= e^{(\pi i/4)s[a]}\vartheta(z;a,b+sa+{}^t(s_1,s_4)),\end{aligned} \tag{4}$$

where we have put

$$s=\begin{pmatrix}s_1 & s_2\\ s_2 & s_4\end{pmatrix},\qquad g=\begin{pmatrix}g_1\\ g_2\end{pmatrix}.$$

Now the new characteristic satisfies

$${}^ta(b+sa+{}^t(s_1,s_4)) \equiv {}^tab + s[a] + (s_1,s_4)a \equiv {}^tab \qquad \bmod 2,$$

and $(a, b+sa+{}^t(s_1,s_4))$ runs over a complete set of representatives modulo 2 if (a,b) varies in the same way. Hence the theta-factors in (3) are permuted up to a certain factor if any translation is applied. The factor in question can be read off from (2) and (4). We obtain

$$\Theta(z+s) = (-1)^{s_1+s_2+s_4}\Theta(z) \tag{5}$$

for any integral symmetric matrix s. Although superfluous, let us compute directly the behavior of Θ with respect to a modular substitution $z \mapsto z[{}^tu]$, where u is unimodular. We infer from the definition that

$$\vartheta(z[{}^tu];a,b) = \vartheta(z;{}^tua,u^{-1}b), \tag{6}$$

and, if (a,b) runs over a complete set of representatives modulo 2 such that ${}^tab$ is even, we have the same situation for the characteristic $({}^tua, u^{-1}b)$. Hence we get

$$\Theta(z[{}^tu]) = \pm\Theta(z),$$

and looking more carefully at (1) and (6) in particular

$$\Theta(z[u]) = -\Theta(z) \qquad \text{for} \quad u=\begin{pmatrix}1 & 0\\ 0 & -1\end{pmatrix}. \tag{7}$$

Finally using the inversion formula for theta-functions we obtain:

$$\begin{aligned}\vartheta(z;a,b) &= \sum_g e^{\pi i(z[g+(a/2)]+{}^tbg)}\\ &= e^{(\pi i/4)z[a]}\sum_g e^{\pi i(z[g]+{}^t(b+za)g)}\\ &= e^{(\pi i/4)z[a]}\det(iz)^{-1/2}\sum_g e^{-\pi iz^{-1}[g+(za+b)/2]}\\ &= e^{-(\pi i/2){}^tab}\det(iz)^{-1/2}\sum_g e^{\pi i(-z^{-1}[g+b/2]-{}^tag)}\\ &= e^{-(\pi i/2){}^tab}\det(iz)^{-1/2}\vartheta(-z^{-1};b,a).\end{aligned}$$

Here $\det(iz)^{-1/2}$ is that branch of the algebraic function which is positive for pure imaginary z. Take the product over the ten characteristics (2). Then the contribution of the factors $e^{-(\pi i/2)^t ab}$ amounts to -1 and that of the factors $\det(iz)^{-1/2}$ to $-\det(z)^{-5}$. Hence we obtain

$$\Theta(-z^{-1}) = \det(-z)^5 \Theta(z).$$

Having found the transformation formulas for a set of generators of Γ_2 we may now state

$$\Theta(m\langle z\rangle) = v(m) j(m, z)^5 \Theta(z)$$

for arbitrary $m \in \Gamma_2$ and $v(m) = \pm 1$. Note that $\vartheta(z; a, b)$ does not vanish identically if ${}^t ab \equiv 0 \bmod 2$, as can be verified easily from the definition of $\vartheta(z; a, b)$ by its Fourier series. So $\Theta(z) \not\equiv 0$ and

$$v: \Gamma_2 \to \{\pm 1\}$$

turns out to be a character of Γ_2. This character is non-trivial by (5). Hence we have found incidentally the only non-trivial multiplier system for $n = 2$ mentioned in §4. It wouldn't be too difficult to determine $v(m)$ more explicitly from the transformation formulas of the $\vartheta(z; a, b)$. We only call attention to

$$v(m) = 1 \quad \text{for} \quad m\langle z\rangle = z[u], \quad u = \begin{pmatrix} 1 & 0 \\ 0 & -1 \end{pmatrix}, \tag{8}$$

which follows from (7). Our results are summarized in

Proposition 1

$\Theta(z)$ is a modular form of weight 5 with a non-trivial character as multiplier system.

Next we determine the zero set

$$V(\Theta) = \{z \in H_n \mid \Theta(z) = 0\}$$

of the function Θ. Let us use the notation

$$z = \begin{pmatrix} z_1 & z_2 \\ z_2 & z_4 \end{pmatrix}$$

and similarly for other two-rowed symmetric matrices. It follows from (7) that $\Theta(z)$ vanishes if $z_2 = 0$. There are no further zeros modulo Γ_2; this is the content of

Proposition 2

The zero set of $\Theta(z)$ is

$$V(\Theta) = \bigcup_{m \in \Gamma_2} m\langle N\rangle, \qquad N = \{z \in H_2 \mid z_2 = 0\}.$$

All the zeros are of order one.

Proof
Consider the subset

$$B = \{z \in H_2 | 0 \leq 2y_2 \leq y_1 \leq y_4, y_1 \geq \sqrt{3}/2, |x_\nu| \leq 1/2 \ (\nu = 1,\dots,4)\}$$

of H_2 which contains Siegel's fundamental domain. Note that the factors of automorphy $j(m,z)^k$ are units in the local rings of holomorphic functions. Hence it is sufficient to study the restriction of Θ on B and to prove its zero set to be

$$V(\Theta|_B) = B \cap N,$$

each zero being of order one. For this purpose we have to investigate the ten factors $\vartheta(z;a,b)$ and their zeros on B separately.
(i) $\vartheta(z;0,b) \neq 0$ for any $z \in B$: By Definition 1

$$\vartheta(z;0,b) = \sum_g (-1)^{{}^t b g} e^{\pi i z[g]}.$$

Subtract 1 for $g = 0$ and take the absolute values of the individual terms in the remaining series. Then we obtain

$$|\vartheta(z;0,b) - 1| \leq \sum_{g \neq 0} e^{-\pi y[g]}.$$

Since $z \in B$ implies $y \geq \frac{\sqrt{3}}{4} 1$ we may continue this estimate by

$$|\vartheta(z;0,b) - 1| \leq 2e^{-\pi y_1} + 2e^{-\pi y_4} + \sum_{g_1^2+g_2^2>1} e^{-\pi(\sqrt{3}/4)(g_1^2+g_2^2)}$$
$$\leq 4e^{-\pi\sqrt{3}/2} + \sum_{g_1^2+g_2^2>1} e^{-\pi(\sqrt{3}/4)(g_1^2+g_2^2)}.$$

It can be shown by numerical inspection that the right-hand side is less than one. Hence $\vartheta(z;0,b)$ cannot vanish on B.
(ii) $\vartheta(z;\binom{1}{0},b) \neq 0$ for any $z \in B$: By Definition 1

$$\vartheta(z;\tbinom{1}{0},b)e^{-(\pi i/4)z_1} = \sum_g (-1)^{{}^t b g} e^{\pi i(z_1 g_1(g_1+1) + z_2 g_2(2g_1+1) + z_4 g_2^2)}.$$

Subtract from this series the terms with $g = 0, \begin{pmatrix} -1 \\ 0 \end{pmatrix}$ and take the absolute values of the individual terms in the remaining series. Then one obtains

$$|\vartheta(z;\tbinom{1}{0},b)e^{-(\pi i/4)z_1} - 2| \leq \sum_{g \neq 0, \binom{-1}{0}} e^{-\pi(y_1 g_1(g_1+1) + y_2 g_2(2g_1+1) + y_4 g_2^2)}.$$

Use the identity

$$g_2(2g_1 + 1) = (g_1 + g_2 + 1)(g_1 + g_2) - g_1(g_1 + 1) - g_2^2$$

to estimate the exponent for any $z \in B$ by

$$\begin{aligned} & y_1 g_1(g_1+1) + y_2 g_2(2g_1+1) + y_4 g_2^2 \\ & \quad \geq (y_1 - y_2) g_1(g_1+1) + (y_4 - y_2) g_2^2 \\ & \quad \geq \frac{\sqrt{3}}{4}(g_1(g_1+1) + g_2^2). \end{aligned}$$

So we get

$$|\vartheta(z; (\begin{smallmatrix}1\\0\end{smallmatrix}), b) e^{-(\pi i/4)z_1} - 2| \leq \sum_{g \neq 0, (\begin{smallmatrix}-1\\0\end{smallmatrix})} e^{-(\pi\sqrt{3}/4)(g_1(g_1+1)+g_2^2)}.$$

Again by numerical inspection the right-hand side turns out to be less than two. Hence $\vartheta(z; (\begin{smallmatrix}1\\0\end{smallmatrix}), b)$ cannot vanish on B.
(iii) $\vartheta(z; (\begin{smallmatrix}0\\1\end{smallmatrix}), b) \neq 0$ for any $z \in B$: Apply the modular substitution

$$z \mapsto z[u], \qquad u = \begin{pmatrix} 0 & 1 \\ 1 & 0 \end{pmatrix}.$$

Then the present case is reduced to the previous one by (6). Note in this connection that we have used only

$$2y_2 \leq y_1, y_4, \qquad \frac{\sqrt{3}}{2} \leq y_1, y_4$$

(and not $z \in B$!) for the proof of (ii), and these conditions are invariant with respect to the modular substitution mentioned above.
(iv) $\vartheta(z; a, b)$ for $a = (\begin{smallmatrix}1\\1\end{smallmatrix})$, $b = \varepsilon(\begin{smallmatrix}1\\1\end{smallmatrix})$, $\varepsilon = 0, 1$: By Definition 1 we have

$$\begin{aligned} & \vartheta(z; a, b) e^{-(\pi i/4)z[a] + \pi i z_2} \\ & \quad = \sum_{g_1, g_2 \in \mathbb{Z}} (-1)^{\varepsilon(g_1+g_2)} e^{\pi i(g_1(g_1+1)(z_1-z_2) + g_2(g_2+1)(z_4-z_2) + (g_1+g_2+1)^2 z_2)} \\ & \quad = \sum_{g_1, g_2 \geq 0} \cdot/\cdot + \sum_{g_1, g_2 < 0} \cdot/\cdot + \sum_{g_1 < 0, g_2 \geq 0} \cdot/\cdot + \sum_{g_1 \geq 0, g_2 < 0} \cdot/\cdot . \end{aligned}$$

The second sum is transformed into the first one by the replacement

$$g_1 \mapsto -g_1 - 1, \qquad g_2 \mapsto -g_2 - 1,$$

and the remaining two sums are both transformed into

$$(-1)^{\varepsilon} \sum_{g_1, g_2 \geq 0} (-1)^{\varepsilon(g_1+g_2)} e^{\pi i(g_1(g_1+1)(z_1-z_2) + g_2(g_2+1)(z_4-z_2) + (g_1-g_2)^2 z_2)}$$

by $g_1 \mapsto -g_1 - 1$ with g_2 fixed, respectively $g_2 \mapsto -g_2 - 1$ with g_1 fixed. So we obtain

$$\begin{aligned} & \vartheta(z; a, b) e^{-(\pi i/4)z[a] + \pi i z_2} \\ & \quad = 2 \sum_{g_1, g_2 \geq 0} \{e^{\pi i(g_1+g_2+1)^2 z_2} + (-1)^{\varepsilon} e^{\pi i(g_1-g_2)^2 z_2}\} \\ & \qquad \times (-1)^{\varepsilon(g_1+g_2)} e^{\pi i(g_1(g_1+1)(z_1-z_2) + g_2(g_2+1)(z_4-z_2))}. \end{aligned}$$

The crucial point is that we can divide the term in the brackets $\{\cdot/.\}$ by $1+(-1)^{\varepsilon}e^{\pi i z_2}$,

$$\{\cdot/.\} = (1+(-1)^{\varepsilon}e^{\pi i z_2})(-1)^{\varepsilon}e^{\pi i(g_1-g_2)^2 z_2} \sum_{n=0}^{(2g_1+1)(2g_2+1)-1} (-1)^{n(\varepsilon-1)}e^{\pi i n z_2}.$$

Then we get

$$\vartheta(z;a,b)e^{-(\pi i/4)z[a]+\pi i z_2} = 2(1+(-1)^{\varepsilon}e^{\pi i z_2})(-1)^{\varepsilon}f(z), \tag{9}$$

where

$$f(z) = \sum_{g_1,g_2\geq 0} (-1)^{\varepsilon(g_1+g_2)}e^{\pi i(g_1(g_1+1)(z_1-z_2)+g_2(g_2+1)(z_4-z_2)+(g_1-g_2)^2 z_2)}$$
$$\times \sum_{n=0}^{(2g_1+1)(2g_2+1)-1} (-1)^{n(\varepsilon-1)}e^{\pi i n z_2}.$$

Now it can be shown that f does not vanish on B. For this purpose subtract the term 1 for $g_1 = g_2 = 0$ and take the absolute values of the individual terms in the remaining series. Then, since $z \in B$, we may use

$$y_1 - y_2 \geq \frac{\sqrt{3}}{4}, \qquad y_4 - y_2 \geq \frac{\sqrt{3}}{4}, \qquad y_2 \geq 0$$

and obtain

$$|f(z)-1| \leq -1 + \left(\sum_{n=0}^{\infty} e^{-(\pi/4)\sqrt{3}n(n+1)}(2n+1)\right)^2.$$

The expression on the right is less than one by numerical inspection, hence $f(z) \neq 0$ everywhere on B. Thus by (9) any zeros of $\vartheta(z;a,b)$ on B are those of the factor

$$1+(-1)^{\varepsilon}e^{\pi i z_2}.$$

So we have zeros of order one on $B \cap N$ if $\varepsilon = 1$ and none if $\varepsilon = 0$. The proposition follows from these results on the ϑ-factors of $\Theta(z)$.

Remarks

(i) In fact we have proved that $\Theta(z)/z_2$ is holomorphic and different from zero everywhere on B, i.e. a global unit on Siegel's fundamental domain.

(ii) It follows from Propositions 1 and 2 that Θ^2 is a non-trivial cusp form of weight 10.

The last proposition has important consequences. Let f be any modular form of even weight which vanishes on N. Then by the proposition f/Θ is holomorphic first on B and, since B contains Siegel's fundamental domain, also on all of H_2. Hence f/Θ is a modular form of odd weight with the non-trivial multiplier system v(m). But it follows from (8) and the transformation formula with respect to the modular subsitution

$$z \mapsto z[u], \qquad u = \begin{pmatrix} 1 & 0 \\ 0 & -1 \end{pmatrix},$$

that any such modular form must vanish on N. Hence by the proposition we may divide by Θ once again and obtain a modular form f/Θ^2, the weight of which has decreased by 10. So we have the

Corollary

Any modular form of even weight which vanishes on N is divisible by Θ^2 in the ring of holomorphic functions on H_2.

Now we turn to the structure theorem for the graded ring

$$M_2^* = \sum_{k \in \mathbb{Z}} \oplus M_2^{2k}$$

of all modular forms of even weights and degree two. Let us first quote the classical result in the one-variable case. Denote the Eisenstein series of the lowest weights 4 and 6 by e_4 and e_6. The connection with the elliptic invariants is given by

$$g_2 = 60\zeta(4)e_4, \qquad g_3 = 140\zeta(6)e_6,$$

where ζ denotes Riemann's ζ-function. Then the graded ring of elliptic modular forms is generated by e_4 and e_6 and these two modular forms are algebraically independent. This means that any elliptic modular form can be represented uniquely as an isobaric polynomial in e_4 and e_6. We want to find an analogue for the graded ring M_2^*.

In the following we always assume the weight k to be even and positive. First we consider the restriction of any modular form $F \in M_2^k$ to the submanifold

$$N = \{z \in H_2 | z_2 = 0\}$$

which is isomorphic to $H_1 \times H_1$.

Proposition 3

Let F be any modular form of degree 2 and positive even weight. Then

$$f(z_1, z_4) = F\begin{pmatrix} z_1 & 0 \\ 0 & z_4 \end{pmatrix} \qquad (z_1, z_4 \in H_1)$$

can be represented by an isobaric polynomial in the three functions

$$e_4(z_1)e_4(z_4), \qquad e_6(z_1)e_6(z_4), \qquad e_4^3(z_1)e_6^2(z_4) + e_4^3(z_4)e_6^2(z_1). \quad (10)$$

Proof

We mean isobaric with respect to the weights 4, 6 and 12. But in the proof we don't have to worry about the polynomial being isobaric owing to an argument already used in connection with Theorem 4.3. The function $f(z_1, z_4)$ is an elliptic modular form in each variable separately if the other variable is kept fixed. Hence

$$f(z_1, z_4) = \sum_{\nu,\mu=1}^{l} a_{\nu\mu} h_\nu(z_1) h_\mu(z_4),$$

where h_ν ($\nu = 1, \ldots, l$) is any basis of the linear space of elliptic modular forms of weight k. Since k is even, F is invariant with respect to the modular substitution

$$z \mapsto z[u], \qquad u = \begin{pmatrix} 0 & 1 \\ 1 & 0 \end{pmatrix}.$$

Therefore $f(z_1, z_4)$ turns out to be symmetric and we get

$$a_{\nu\mu} = a_{\mu\nu} \qquad (\mu, \nu = 1, \ldots, l)$$

for the coefficients. Take in particular the basis

$$e_4^{\alpha_\nu} e_6^{\beta_\nu}, \qquad 4\alpha_\nu + 6\beta_\nu = k, \qquad \alpha_\nu, \beta_\nu \geq 0;$$

then we obtain the representation

$$f(z_1, z_4) = \sum_{\nu \leq \mu} a_{\nu\mu} (e_4^{\alpha_\nu}(z_1) e_6^{\beta_\nu}(z_1) e_4^{\alpha_\mu}(z_4) e_6^{\beta_\mu}(z_4) + e_4^{\alpha_\nu}(z_4) e_6^{\beta_\nu}(z_4) e_4^{\alpha_\mu}(z_1) e_6^{\beta_\mu}(z_1)).$$

We factor out the term

$$(e_4(z_1) e_4(z_4))^{\min(\alpha_\nu, \alpha_\mu)} (e_6(z_1) e_6(z_4))^{\min(\beta_\nu, \beta_\mu)}$$

and observe that $4\alpha_\nu + 6\beta_\nu = k$. Then $f(z_1, z_4)$ becomes a polynomial in

$$e_4(z_1) e_4(z_4), \quad e_6(z_1) e_6(z_4), \quad e_4^\alpha(z_1) e_6^\beta(z_4) + e_4^\alpha(z_4) e_6^\beta(z_1),$$

where α and β are non-negative and satisfy $2\alpha = 3\beta$. So we only have to reduce the exponents α and β to the lowest possible values $\alpha = 3$, $\beta = 2$. Because $2\alpha = 3\beta$ the number $h = \alpha/3 = \beta/2$ is integral and we may write

$$e_4^\alpha(z_1) e_6^\beta(z_4) + e_4^\alpha(z_4) e_6^\beta(z_1) = (e_4^3(z_1) e_6^2(z_4))^h + (e_4^3(z_4) e_6^2(z_1))^h.$$

Now by induction on h and a straightforward calculation it is easily verified that

$$(e_4^3(z_1) e_6^2(z_4))^h + (e_4^3(z_4) e_6^2(z_1))^h$$

is indeed a polynomial in

$$e_4(z_1) e_4(z_4), \qquad e_6(z_1) e_6(z_4), \qquad e_4^3(z_1) e_6^2(z_4) + e_4^3(z_4) e_6^2(z_1).$$

Next we show that the three functions (10) can be realized as restrictions of modular forms of degree two on the submanifold N. Denote Siegel's Eisenstein series of degree two and weight k by E_k ($E^k_{2,0}$ in the former notation of §5). The restrictions $E_k|_N$ are isobaric polynomials in the three functions (10) by Proposition 3, and because of the weights we obtain in particular

$$E_4\begin{pmatrix} z_1 & 0 \\ 0 & z_4 \end{pmatrix} = a\, e_4(z_1)e_4(z_4), \qquad E_6\begin{pmatrix} z_1 & 0 \\ 0 & z_4 \end{pmatrix} = b\, e_6(z_1)e_6(z_4).$$

Since the constant terms in the Fourier expansions of Eisenstein series are one, we get $a = b = 1$. Hence the restrictions of E_4 and E_6 to N yield the first two functions (10). Concerning the third one we get from the previous proposition

$$E_{12}\begin{pmatrix} z_1 & 0 \\ 0 & z_4 \end{pmatrix} = c_1 E_4^3\begin{pmatrix} z_1 & 0 \\ 0 & z_4 \end{pmatrix} + c_2 E_6^2\begin{pmatrix} z_1 & 0 \\ 0 & z_4 \end{pmatrix}$$
$$+ c_3\{e_4^3(z_1)e_6^2(z_4) + e_4^3(z_4)e_6^2(z_1)\}$$

with certain constants c_1, c_2 and c_3. If c_3 were zero, $E_{12} - c_1 E_4^3 - c_2 E_6^2$ would vanish on N; hence, by the corollary of Proposition 2,

$$(E_{12} - c_1 E_4^3 - c_2 E_6^2)/\Theta^2$$

would become a modular form of weight 2. But such a modular form vanishes on N by Proposition 3 and is then again divisible by Θ^2. The result would be a modular form of negative weight, hence zero. So finally we would get

$$E_{12} = c_1 E_4^3 + c_2 E_6^2. \tag{11}$$

We show the absurdity of such a relation below and then $c_3 \neq 0$ becomes available. So the restriction of the modular form

$$E_{12} - c_1 E_4^3 - c_2 E_6^2$$

to N yields the remaining third function in (10) up to a constant factor.

Having found 'liftings' of the three functions (10) to modular forms of degree two, we derive from Proposition 3 that the restriction of any modular form f of even weight and degree two onto N can be represented as

$$f|_N = p(E_4|_N, E_6|_N, E_{12}|_N)$$

with an appropriate polynomial p. Hence,

$$f - p(E_4, E_6, E_{12})$$

vanishes on N. By the corollary of Proposition 2 we may divide by Θ^2 and can repeat our argument. With each step the weight decreases by 10, so finally we arrive at the zero-function. Summarizing, we have proved that the graded ring M_2^* is generated by the four modular forms

$$E_4, E_6, E_{12} \text{ and } \Theta^2. \tag{12}$$

As a supplement we show the absurdity of (11). Unfortunately there are no analytical arguments which allow us to recognize the impossibility of such a relation. Hence we are left with comparing Fourier expansions. The computation of the Fourier coefficients is difficult and tedious, so we take the following values from [34]:

$$E_4\colon a\begin{pmatrix}0&0\\0&0\end{pmatrix}=1,\ a\begin{pmatrix}1&0\\0&0\end{pmatrix}=2^4\cdot3\cdot5,\quad a\begin{pmatrix}1&0\\0&1\end{pmatrix}=2^5\cdot3^3\cdot5\cdot7,$$

$$E_6\colon a\begin{pmatrix}0&0\\0&0\end{pmatrix}=1,\ a\begin{pmatrix}1&0\\0&0\end{pmatrix}=-2^3\cdot3^2\cdot7,\ a\begin{pmatrix}1&0\\0&1\end{pmatrix}=2^4\cdot3^3\cdot5\cdot7\cdot11,$$

$$E_{12}\colon a\begin{pmatrix}0&0\\0&0\end{pmatrix}=1,\quad a\begin{pmatrix}1&0\\0&0\end{pmatrix}=\frac{2^4\cdot3^2\cdot5\cdot7\cdot13}{691},$$

$$a\begin{pmatrix}1&0\\0&1\end{pmatrix}=\frac{2^5\cdot3^3\cdot5\cdot7\cdot13\cdot19\cdot23\cdot2659}{131\cdot593\cdot691}.$$

The comparison of the first two Fourier coefficients in (11) yields

$$1=c_1+c_2,\qquad \frac{2^4\cdot3^2\cdot5\cdot7\cdot13}{691}=2^4\cdot3^2\cdot5c_1-2^4\cdot3^2\cdot7c_2,$$

which implies

$$c_1=\frac{7\cdot691+5\cdot7\cdot13}{691\cdot12},\qquad c_2=\frac{5\cdot691-5\cdot7\cdot13}{691\cdot12}. \tag{13}$$

The comparison of the third Fourier coefficients in (11) demands

$$\frac{2^5\cdot3^3\cdot5\cdot7\cdot13\cdot19\cdot23\cdot2659}{131\cdot593\cdot691}$$
$$=(2^5\cdot3^4\cdot5\cdot7+2^9\cdot3^3\cdot5^2)c_1+(2^5\cdot3^3\cdot5\cdot7\cdot11+2^7\cdot3^4\cdot7^2)c_2,$$

which is incompatible with (13), since for instance the prime 131 does not appear in the prime number decomposition of the right-hand side.

Finally we want a list of generators consisting only of Eisenstein series. By our former results we have

$$E_{10}=a_1E_4E_6+a_2\Theta^2.$$

Comparing Fourier coefficients it can be shown that a_2 is different from zero. Hence we may replace Θ^2 by E_{10} in the list of generators (12). So the following main result of this section is proved.

Theorem

The graded ring of modular forms of even weights and degree two is generated as a $\mathbb{C}$-algebra by the four Eisenstein series of weights 4, 6, 10 *and* 12.

In the course of §10 we will see that these generators are algebraically independent. Hence the representation of modular forms as polynomials in E_4, E_6, E_{10} and E_{12} is unique. From this fact we deduce at once the following

Corollary

For even k

$$\dim M_2^k = \#\{(v_1, \ldots, v_4) \in \mathbb{Z}^4 \mid v_1, \ldots, v_4 \geq 0, 4v_1 + 6v_2 + 10v_3 + 12v_4 = k\}.$$

Remark

For $k < 10$ we have $\dim M_2^k \leq 1$, and $\dim M_2^{10} = 2$. Since Eisenstein series exist for $k > 2$, we haven't any non-trivial cusp forms of even weight $k < 10$. On the other hand, Θ^2 is a cusp form of weight 10. Hence Θ^2 is characterized up to a constant factor as the cusp form of lowest possible even weight.

V

Modular functions

10 Quotients of modular forms

It is certainly not satisfactory if we define modular functions in the narrow sense to be just quotients of modular forms of equal weight. Nevertheless let us first take this point of view because of the advantage of being able to apply our knowledge of modular forms quite directly. In §11 we shall discuss how this conception can be weakened; so, finally, a comprehensive and satisfactory theory will arise.

Definition

For any positive integer n let Q_n be the following field of meromorphic functions on H_n:

$$Q_n = \left\{ f = \frac{g}{h} \mid g, h \text{ modular forms of equal weight}, h \neq 0 \right\}.$$

The elements of Q_n are called modular functions of degree n 'in the narrow sense'.

Obviously any $f \in Q_n$ is invariant with respect to the action of the modular group. For the one-variable case it is well known that the elements of Q_1 can also be characterized by the conditions of being meromorphic on $H_1 \cup \{\infty\}$ and invariant with respect to the elliptic modular group. We will follow up the corresponding question for $n > 1$ later.

First we consider algebraic relations over $\mathbb{C}$ between modular forms on the one hand and between elements of Q_n on the other. For any fixed l, each family of $l + 1$ modular forms is algebraically dependent if and only if each family of l functions in Q_n is algebraically dependent. Indeed, assume the algebraic dependence of each family of $l + 1$ modular forms and take l elements $f_1, \ldots, f_l$ of Q_n. Write these elements

$$f_\nu = \frac{g_\nu}{g_0} \qquad (\nu = 1, \ldots, l)$$

as quotients of modular forms of equal weight k and with common denominator g_0. Then the $l + 1$ modular forms g_ν $(\nu = 0, \ldots, l)$ satisfy a non-trivial algebraic homogeneous equation

$$\sum_{\nu_0+\cdots+\nu_l=h} c_\nu g_0^{\nu_0}\cdots g_l^{\nu_l}=0.$$

Dividing by g_0^h yields an algebraic equation

$$\sum_\nu c_\nu f_1^{\nu_1}\cdots f_l^{\nu_l}=0$$

for $f_1,\ldots,f_l$. Conversely, assume the algebraic dependence of each family of l modular functions in Q_n and consider $l+1$ modular forms $g_0,\ldots,g_l$ of weights $k_0,\ldots,k_l$. We may suppose without loss of generality that the weights are positive and $g_0\neq 0$. Set $k=k_0k_1\ldots k_l$ and form the l modular functions

$$\frac{g_1^{k/k_1}}{g_0^{k/k_0}},\ldots,\frac{g_l^{k/k_l}}{g_0^{k/k_0}}.$$

These functions satisfy a non-trivial algebraic equation. Multiply this equation by a suitable power of g_0 in order to remove the denominators. Then we obtain an algebraic equation for $g_0,\ldots,g_l$.

We have proved in Theorem 4.3 that each family of $n(n+1)/2+2$ modular forms is algebrically dependent. Hence each family of $n(n+1)/2+1$ modular functions in the narrow sense satisfies a non-trivial algebraic equation over the complex numbers. Beyond this we have proved a more detailed statement about the degree of the algebraic equation between modular forms. From there we get

Proposition 1

Let $f_1,\ldots,f_s$ be s algebraically independent elements of Q_n, $s=n(n+1)/2$. Then any $f\in Q_n$ satisfies an algebraic equation

$$p(f,f_1,\ldots,f_s)=0$$

of bounded degree with respect to f.

Before proving this let us make the following remark. By the proposition any $f\in Q_n$ is algebraic over the rational function field $\mathbb{C}(f_1,\ldots,f_s)$. The degree of f over $\mathbb{C}(f_1,\ldots,f_s)$ certainly depends on the choice of $f_1,\ldots,f_s$, but, as the proposition states, it is bounded independently of f if $f_1,\ldots,f_s$ are fixed. The existence of s algebraically independent elements of Q_n is still unproved. But supposing their existence, we would obtain that Q_n is a finite algebraic extension of $\mathbb{C}(f_1,\ldots,f_s)$ generated by an element f of maximal degree. Hence, anticipating the existence of s algebraically independent modular functions in the narrow sense we have as a corollary:

Corollary

The field Q_n is an algebraic function field of transcendence degree $n(n+1)/2$ over $\mathbb{C}$.

Proof

Choose representations of $f_1, \ldots, f_s$ as quotients of modular forms

$$f_\nu = \frac{g_\nu}{g_0} \qquad (\nu = 1, \ldots, s)$$

with common denominator. Then all modular forms $g_0, \ldots, g_s$ have the same weight, say l. We now apply the full statement of Theorem 4.3. For any

$$f = \frac{\xi}{\eta}$$

in Q_n, represented as the quotient of modular forms ξ and η of weight k, we get algebraic equations

$$\sum c\xi^\nu g_0^{\nu_0} \ldots g_s^{\nu_s} = 0, \qquad \sum d\eta^\nu g_0^{\nu_0} \ldots g_s^{\nu_s} = 0,$$

where the exponents satisfy

$$\nu k + (\nu_0 + \cdots + \nu_s) l = \mu k l^{s+1},$$

μ depending only on n. Hence the degrees with respect to ξ and η are bounded by μl^{s+1}, which is independent of f. On eliminating ξ and η from these two algebraic equations we obtain an algebraic equation for f of bounded degree over $\mathbb{C}(f_1, \ldots, f_s)$.

Now we turn to the existence of $n(n+1)/2$ algebraically independent modular functions in Q_n. The proof does not use any specific properties of the modular group but works for arbitrary discontinuous groups acting on a bounded domain as a group of biholomorphic automorphisms. In our case we take the generalized unit-circle D_n of degree n as the underlying domain and use the notations of §§1 and 3.

Proposition 2

Let Γ be any subgroup of Φ_n operating discontinuously on the domain D_n. Then there exist $s = n(n+1)/2$ analytically independent automorphic functions with respect to Γ which are quotients of automorphic forms.

Proof

We may assume $\pm 1 \in \Gamma$. Denote by Γ_0 the finite subgroup of the volume-preserving elements of Γ and write

$$j_m(w) = \det\left(\frac{\partial m\langle w\rangle}{\partial w}\right)$$

for the Jacobian of any map m in Φ_n. In §3 we considered the subset

$$B = \{w \in D_n \,|\, |j_m(w)| \leq 1 \text{ for all } m \in \Gamma\}$$

of D_n, which turned out to be more or less a fundamental domain of the group Γ. Choose any interior point $w_0 = (w_{\mu\nu}^{(0)})$ of B, which is not a fixed point of Γ. Then by Theorem 3.4 we have

$$|j_m(w)| \begin{cases} <1 & \text{for all } m \in \Gamma - \Gamma_0 \\ =1 & \text{for } m \in \Gamma_0 \end{cases}$$

and w belonging to a sufficiently small neighborhood $U(w_0)$ of w_0. Now consider the Poincaré series

$$P_{lr}(w; f) = \sum_{m \in \pm 1 \backslash \Gamma} f(m\langle w \rangle) j_m(w)^{lr},$$

where f is any bounded holomorphic function on D_n and r the order of Γ_0. By Theorem 3.3 we have uniform convergence on compact subsets of D_n. Furthermore

$$j_m^r(w) = 1$$

for all $m \in \Gamma_0$. Hence, passing to the limit $l \to \infty$, we obtain

$$\lim_{l \to \infty} P_{lr}(w; f) = \sum_{m \in \pm 1 \backslash \Gamma_0} f(m\langle w \rangle)$$

uniformly in $w \in U(w_0)$. Take in particular the following functions for f:

$$f_0 = \frac{2}{r}, \qquad f_{\mu\nu}(w) = (w_{\mu\nu} - w_{\mu\nu}^{(0)})\varphi^2(w) \qquad (\mu, \nu = 1, \ldots, n; \mu \leq \nu),$$

where φ is any polynomial with

$$\varphi(w_0) = 1, \qquad \varphi(m\langle w_0 \rangle) = 0 \quad \text{for} \quad \pm 1 \neq m \in \Gamma_0.$$

Such polynomials do exist, since w_0 was assumed to be a non-fixed point of Γ. Then the automorphic functions

$$F_{\mu\nu}(w; l) = \frac{P_{lr}(w; f_{\mu\nu})}{P_{lr}(w; f_0)} \qquad (\mu, \nu = 1, \ldots, n; \mu \leq \nu)$$

tend to

$$\sum_{m \in \pm 1 \backslash \Gamma_0} f_{\mu\nu}(m\langle w \rangle) \qquad (\mu, \nu = 1, \ldots, n; \mu \leq \nu) \tag{1}$$

uniformly on $U(w_0)$ for $l \to \infty$. Hence the Jacobian of the functions $F_{\mu\nu}$ $(\mu \leq \nu)$ at the point w_0 tends to the Jacobian of the functions (1) at w_0, which is one by the choice of the polynomial φ. So the Jacobian of the automorphic functions $F_{\mu\nu}(*, l)$ at the point w_0 is certainly different from zero if l is sufficiently large. Hence the $F_{\mu\nu}$ become analytically, and all the more algebraically, independent.

So in particular for the modular group we have completed the proof of the corollary to Proposition 1 saying that the field Q_n of modular functions in the narrow sense is an algebraic function field of transcendence degree $n(n+1)/2$. If we transfer the analytically independent functions constructed in Proposition 2 to Siegel's half-space we obtain modular functions represented as quotients of Poincaré series of type II in the sense of §6, thereby demonstrating the significance of this kind of Poincaré series. Finally we want to point out that the algebraically independent modular functions constructed above are quotients of cusp forms.

Our results in §9 on the graded ring of modular forms of degree two and even weights allow complete insight into the structure of Q_2. Consider the three modular functions

$$\frac{E_4E_6}{E_{10}}, \quad \frac{E_6^2}{E_{12}}, \quad \frac{E_4^5}{E_{10}^2}. \tag{2}$$

By the theorem of §9 any modular function f in Q_2 is the quotient of two isobaric polynomials in E_4, E_6, E_{10} and E_{12} of equal weight. Divide numerator and denominator by any monomial of such a kind; then we obtain a representation of f as quotient of two sums, the terms of which are of the form

$$E_4^{v_1}E_6^{v_2}E_{10}^{v_3}E_{12}^{v_4}, \qquad 4v_1+6v_2+10v_3+12v_4=0,$$

and where the exponents may become negative. From the relation

$$E_4^{v_1}E_6^{v_2}E_{10}^{v_3}E_{12}^{v_4} = \left(\frac{E_4E_6}{E_{10}}\right)^{v_2+2v_4}\left(\frac{E_6^2}{E_{12}}\right)^{-v_4}\left(\frac{E_4^5}{E_{10}^2}\right)^{v_1+v_2+2v_3+2v_4}$$

we deduce a representation of f as a rational function of the three modular functions (2). These three functions are algebraically independent since otherwise the transcendence degree of Q_2 over the complex numbers would be less than two. So we may state

Proposition 3

The field Q_2 is the rational function field generated over $\mathbb{C}$ by the algebraically independent functions

$$\frac{E_4E_6}{E_{10}}, \quad \frac{E_6^2}{E_{12}}, \quad \frac{E_4^5}{E_{10}^2}.$$

Corollary

The generators E_4, E_6, E_{10} and E_{12} of the graded ring M_2^* in §9 are algebraically independent.

This proposition generalizes the well-known result on the generation of the field of elliptic modular functions by the modular invariant $j(z)$. But in general the function fields Q_n are by no means rational. The reader should consult the work of E. Freitag [24], Y.-S. Tai [66] and D. Mumford [53] about this question. The most far-reaching result is due to Mumford: Q_n is certainly non-rational for $n \geq 7$.

11 Pseudoconcavity

A proper treatment would demand a modular function to be a meromorphic function on H_n, invariant with respect to the modular group and meromorphic at infinity in some sense. This conception is suggested by the one-variable case. Note that the last condition is not superfluous for $n = 1$: the function $e^{j(z)}$ is meromorphic on H_1 and invariant but has an essential singularity at infinity. Therefore the efforts of several mathematicians were first focused on finding a compactification of H_n/Γ_n and proving that any meromorphic function on this compactification can be represented globally as a quotient of modular forms. This was realized by I. Satake [60] with remarkable success; other attempts were made by C.L. Siegel [65] and U. Christian [14]. Later a different kind of compactification was investigated by Mumford. But something unexpected happened. Still using Satake's compactification, W.L. Baily discovered in [4] that any condition at 'infinity' is superfluous, provided $n > 1$; i.e., in contrast to the case $n = 1$, for $n > 1$ any meromorphic function on H_n which is invariant with respect to Γ_n is a quotient of modular forms and hence an element of Q_n. From there it became desirable to find an approach to Baily's result avoiding any kind of compactification whatsoever. We present such a procedure in this section; it is due to A. Andreotti and H. Grauert [1].

To begin with we explain the concept of pseudoconcavity. If U is any open subset of the complex number space $\mathbb{C}^n$ and $D \subset U$, then the holomorphically convex hull of D in U is defined as

$$\hat{D}_U = \left\{ z \in U \,\middle|\, |f(z)| \leq \sup_D |f| \text{ for all holomorphic } f \text{ on } U \right\}.$$

This concept is borrowed from the characterization of regions of holomorphy in several complex variables. Clearly we have

$$D \subset \hat{D}_U \subset U.$$

Definition 1

Let A be a subset of $\mathbb{C}^n$ and z a point of its boundary ∂A. Then z is called a pseudoconcave boundary point of A if there exist arbitrarily small open neighborhoods U of z such that z is an interior point of $\widehat{(U \cap A)}_U$.

We mean by 'arbitrarily small' that each neighborhood of z contains a neighborhood of the indicated kind. Next we define the pseudoconcavity of a group.

Definition 2

Let G be any domain in $\mathbb{C}^n$ and Γ a group of biholomorphic automorphisms of G. The group Γ is called pseudoconcave if there exists a non-empty open set A lying relatively compact in G such that any point of its closure $\bar{A}$ can be mapped by Γ either onto a point of A or onto a pseudoconcave boundary point of A.

Examples are given by any discontinuous group Γ acting on a domain G with compact fundamental domain. Indeed, enlarge the fundamental domain a little to obtain an open relatively compact subset A of G. Then A trivially satisfies the requirements of Definition 2, since each point of G is equivalent to a point of A modulo Γ. Another example will be Siegel's modular group Γ_n of degree $n > 1$. This will be the content of a theorem to be proved at the end of this section.

We are now able to formulate the main theorem.

Theorem 1

Let Γ be any pseudoconcave group of biholomorphic automorphisms of a domain G in $\mathbb{C}^n$ and $K(\Gamma)$ the corresponding field of automorphic functions. Assume $f_1, \ldots, f_n$ to be algebraically independent elements of $K(\Gamma)$. Then $K(\Gamma)$ is a finite algebraic extension of the rational field $\mathbb{C}(f_1, \ldots, f_n)$, hence an algebraic function field of transcendence degree n over the complex numbers.

Of course by an automorphic function we mean any meromorphic function on G which is invariant with respect to Γ. As a consequence of the theorem we remark that the elliptic modular group cannot be pseudoconcave since $e^{j(z)}$ is not algebraic over $\mathbb{C}(j(z))$.

Before going into the details of the proof and anticipating the pseudoconcavity of Siegel's modular group Γ_n for $n > 1$, we draw the reader's attention to the most important consequences. First we weaken the concept of a modular function in the narrow sense in the following final and more satisfactory way.

Definition 3

A modular function f of degree n is a meromorphic function on H_n, which is invariant with respect to the action of the modular group. For $n = 1$ we additionally demand f to be meromorphic at infinity.

Then our first corollary puts a deep insight into simple terms.

Corollary (i)

Each modular function is the quotient of two modular forms of equal weight. Hence 'modular functions' and 'modular functions in the narrow sense' are the same.

Proof

The case $n = 1$ is exceptional by definition and well known. Assume $n > 1$ and choose any set of algebraically independent modular functions

$$f_1, \ldots, f_s, \qquad s = n(n+1)/2,$$

in Q_n (cf. Proposition 10.2). Now, if f is any modular function in the sense of Definition 3, then f satisfies an algebraic equation

$$f^l + a_{l-1} f^{l-1} + \cdots + a_0 = 0$$

over $\mathbb{C}(f_1, \ldots, f_s)$ by Theorem 1. Hence the coefficients a_ν are quotients of modular forms, say with a common denominator g. Multiplying this equation by g^l yields an algebraic equation for gf, the coefficients of which are modular forms, the first coefficient being one. Since the ring of holomorphic functions is integrally closed in the field of meromorphic functions, gf turns out to be holomorphic and hence a modular form.

The next corollary offers a different proof for the algebraic dependence of sufficiently many modular forms. It does not use the estimates of $\dim M_n^k$ as in Theorem 4.3.

Corollary (ii)

Any family of $n(n+1)/2 + 2$ modular forms is algebraically dependent over $\mathbb{C}$.

Proof

We saw at the beginning of §10 that this assertion is equivalent to the algebraic dependence of any family of $n(n+1)/2 + 1$ modular functions in the narrow sense. But for $n > 1$ this latter fact is indeed an immediate consequence of Theorem 1.

To prepare for the proof of Theorem 1 we show a generalization of Schwarz's lemma to several complex variables.

Lemma

Let $P_r(a)$ be the open polydisc in $\mathbb{C}^n$ of radius r and center at a. If f is holomorphic on $P_r(a)$ and vanishes of order at least l at a, then

$$|f(z)| \le \left(\frac{|z-a|}{r}\right)^l \sup_{P_r(a)} |f|$$

for all $z \in P_r(a)$.

Proof

We may assume $a = 0$ and $l \ge 1$. Consider for any fixed $0 \ne z \in P_r(0)$ the function

$$f(tz)$$

of the single variable t in $|t| < r/|z|$. This function vanishes of order at least l at $t = 0$. Hence Schwarz's lemma in one variable yields

$$|f(tz)| \le \frac{|t|^l}{r^l}|z|^l \sup_{|t|<r/|z|} |f(tz)|.$$

Set $t = 1$ to obtain

$$|f(z)| \le \left(\frac{|z|}{r}\right)^l \sup_{P_r(0)} |f|.$$

Proof of Theorem 1

Let $f_1, \dots, f_n$ be the given algebraically independent elements of $K(\Gamma)$, g another function of $K(\Gamma)$ and A a non-empty open set lying relatively compact in G and which is relevant for the pseudoconcavity of Γ (cf. Definition 2). Then we assign four objects to any point $z \in \bar{A}$ in this order:

(i) $z \in \bar{A} \mapsto \gamma \in \Gamma$, where γ is chosen such that $\gamma(z) \in A$ or $\gamma(z)$ is a pseudoconcave boundary point of A;

(ii) $z \in \bar{A} \mapsto U(z)$, where $U(z)$ is an open neighborhood of z such that
  (α) $f_1, \dots, f_n, g$ are quotients of everywhere coprime holomorphic functions on $\overline{U(z)}$,
  (β) $\gamma(U(z)) \subset A$ if $\gamma(z) \in A$, and $\gamma(z)$ is an interior point of $\widehat{(\gamma(U) \cap A)}_{\gamma(U)}$ if $\gamma(z)$ is a pseudoconcave boundary point of A;

(iii) $z \in \bar{A} \mapsto V(z)$, where $V(z)$ is an open polydisc with center at z such that $\overline{V(z)} \subset U(z)$ if $\gamma(z) \in A$, and $\gamma(\overline{V(z)}) \subset \widehat{(\gamma(U) \cap A)}_{\gamma(U)}$ if $\gamma(z)$ is a pseudoconcave boundary point of A;

(iv) $z \in \bar{A} \mapsto V^*(z) = \frac{1}{2}V(z)$.

The possibility of such an assignment is obvious from the definitions. Then we have for any $z \in \bar{A}$

$$\overline{V(z)} \subset \widehat{(U(z) \cap \gamma^{-1}(A))}_{U(z)}; \tag{1}$$

indeed

$$\overline{V(z)} \subset U(z) = (\widehat{U(z)})_{U(z)} = (\widehat{U(z) \cap \gamma^{-1}(A)})_{U(z)}$$

if $\gamma(z) \in A$, and

$$\overline{V(z)} \subset \gamma^{-1}(\widehat{\gamma(U) \cap A})_{\gamma(U)} = (\widehat{U \cap \gamma^{-1}(A)})_U$$

if $\gamma(z)$ is a pseudoconcave boundary point of A. Since $\bar{A}$ is compact we can choose finitely many points $z_1, \ldots, z_k \in \bar{A}$ such that

$$\bar{A} \subset \bigcup_{\nu=1}^{k} V^*(z_\nu).$$

Let

$$\{(\gamma_\nu, U_\nu, V_\nu, V_\nu^*) | \nu = 1, \ldots, k\}$$

be the family of objects associated with these points by (i)–(iv). Then the following conditions are satisfied:

(a) $V_\nu^* \subset V_\nu \subset U_\nu$ are open subsets of $\mathbb{C}^n$, V_ν^*, V_ν are polydiscs with center $z_\nu \in \bar{A}_\nu$, $V_\nu^* = \frac{1}{2}V_\nu$, and $\{V_\nu^* | \nu = 1, \ldots, k\}$ is a covering of $\bar{A}$;

(b) $\bar{V}_\nu \subset (\widehat{U_\nu \cap \gamma_\nu^{-1}(A)})_{U_\nu}$ for $\nu = 1, \ldots, k$;

(c) there are local representations

$$g = \frac{g'_\nu}{g''_\nu}, \qquad f_\mu = \frac{f'_{\mu\nu}}{f''_{\mu\nu}} \qquad (\nu = 1, \ldots, k; \mu = 1, \ldots, n)$$

on $\bar{U}_\nu$ by quotients of holomorphic functions;

(d) the functions

$$g_{\nu_1,\nu_2} = \frac{g''_{\nu_1}}{g''_{\nu_2} \circ \gamma_{\nu_1}} \quad \text{and} \quad f_{\nu_1,\nu_2} = \prod_{\mu=1}^{n} \frac{f''_{\mu\nu_1}}{f''_{\mu\nu_2} \circ \gamma_{\nu_1}} \qquad (\nu_1, \nu_2 = 1, \ldots, k)$$

are holomorphic on $U_{\nu_1,\nu_2} = \bar{U}_{\nu_1} \cap \gamma_{\nu_1}^{-1}(\bar{U}_{\nu_2})$.

Indeed, these conditions are immediate consequences of (i)–(iv); in particular (b) follows from (1), and (d) because of the invariance of $g, f_1, \ldots, f_n$ with respect to Γ and condition (ii). Let us add an important remark. We did use the coprimality of the local representations of $f_1, \ldots, f_n, g$ by quotients of holomorphic functions according to (ii) in order to prove (d). But from now on we no longer demand coprimality in (c). Choose M, $M' \geq 1$ such that

$$\sup_{U_{\nu_1,\nu_2}} |g_{\nu_1,\nu_2}| < M', \quad \sup_{U_{\nu_1,\nu_2}} |f_{\nu_1,\nu_2}| < M \quad (\nu_1, \nu_2 = 1, \ldots, k). \tag{2}$$

We consider for $s, t, l = 0, 1, 2, \ldots$ the following vector spaces,

$$T(s,t,l) = \left\{ h \in K(\Gamma) | h = \frac{h_\nu}{g_\nu''^s(\prod_\mu f_{\mu\nu}'')^t} \quad \text{on } \bar{U}_\nu, \quad 1 \leq \nu \leq k \right\},$$

where h_ν is holomorphic on $\bar{U}_\nu$ and vanishes at least of order l in z_ν. We want to estimate the dimension of these vector spaces. First by Schwarz's lemma we have

$$\sup(|h_\nu(z)|; z \in \bar{V}_\nu^*, 1 \leq \nu \leq k) \leq \frac{1}{2^l} \sup(|h_\nu(z)|; z \in \bar{V}_\nu, 1 \leq \nu \leq k). \quad (3)$$

On the other hand we shall prove for any $h \in T(s,t,l)$

$$\sup(|h_\nu(z)|; z \in \bar{V}_\nu, 1 \leq \nu \leq k) \leq M'^s M^t \sup(|h_\nu(z)|; z \in \bar{V}_\nu^*, 1 \leq \nu \leq k). \quad (4)$$

Indeed, since $h \in K(\Gamma)$, we have on U_{ν_1,ν_2}

$$\frac{h_{\nu_1}}{g_{\nu_1}''^s(\prod_\mu f_{\mu\nu_1}'')^t} = \frac{h_{\nu_2} \circ \gamma_{\nu_1}}{(g_{\nu_2}'' \circ \gamma_{\nu_1})^s(\prod_\mu f_{\mu\nu_2}'' \circ \gamma_{\nu_1})^t}$$

or

$$h_{\nu_1} = (h_{\nu_2} \circ \gamma_{\nu_1}) g_{\nu_1,\nu_2}^s f_{\nu_1,\nu_2}^t.$$

Since $U_{\nu_1} \cap \gamma_{\nu_1}^{-1}(A \cap V_{\nu_2}^*) \subset U_{\nu_1,\nu_2}$ we get from (2)

$$\sup_{U_{\nu_1} \cap \gamma_{\nu_1}^{-1}(A \cap V_{\nu_2}^*)} |h_{\nu_1}| \leq M'^s M^t \sup(|h_\nu(z)|; z \in \bar{V}_\nu^*, 1 \leq \nu \leq k).$$

The V_ν^* ($\nu = 1, \ldots, k$) cover A, hence

$$\sup(|h_\nu(z)|; z \in U_\nu \cap \gamma_\nu^{-1}(A), 1 \leq \nu \leq k)$$
$$\leq M'^s M^t \sup(|h_\nu(z)|; z \in \bar{V}_\nu^*, 1 \leq \nu \leq k).$$

By the definition of the holomorphically convex hull the left-hand side does not change if $U_\nu \cap \gamma_\nu^{-1}(A)$ is replaced by

$$(\widehat{U_\nu \cap \gamma_\nu^{-1}(A)})_{U_\nu},$$

and since $\bar{V}_\nu$ is contained in this holomorphically convex hull by (b), we finally get (4). Next we combine (3) and (4) to infer

$$2^l \leq M'^s M^t \quad (5)$$

if $T(s,t,l) \neq 0$. Set

$$a = \left[\frac{\log M'}{\log 2}\right] + 1, \qquad b = \left[\frac{\log M}{\log 2}\right] + 1.$$

Then $l \geq sa + tb$ and $(s,t) \neq 0$ imply $T(s,t,l) = 0$. From there we get the estimate

$$\dim T(s,t,0) \leq k\binom{n + sa + tb - 1}{n} \qquad (s,t) \neq 0, \tag{6}$$

for otherwise there would be a function different from zero in $T(s,t,0)$ with zeros of order $sa + tb$ in z_ν $(\nu = 1,\dots,k)$ and such a function would belong to $T(s,t,sa + tb)$. The right-hand side of (6) is a polynomial in t of degree n with leading coefficient $kb^n/n!$. Consider the functions

$$g^{m_0} f_1^{m_1} \dots f_n^{m_n}, \qquad 0 \leq m_0 \leq s, \qquad m_1 + \cdots + m_n \leq t \tag{7}$$

which certainly belong to $T(s,t,0)$. Their number is

$$(s+1)\binom{n+t}{n} = \frac{s+1}{n!}t^n + \cdots.$$

Hence if we choose

$$s = kb^n \tag{8}$$

and then t sufficiently large, the number of functions in (7) exceeds the dimension of $T(s,t,0)$. So $g, f_1, \dots, f_n$ become algebraically dependent and we have proved that $K(\Gamma)$ is algebraic over $\mathbb{C}(f_1,\dots,f_n)$.

We are not yet finished with the proof, since the theorem states that $K(\Gamma)$ is a finite algebraic extension field of $\mathbb{C}(f_1,\dots,f_n)$. During the proof we obtained a bound for the degree of g over $\mathbb{C}(f_1,\dots,f_n)$ by (8). This bound depends on k and b and therefore on g via the local representations (c) of g and the covering U_ν $(\nu = 1,\dots,k)$. To obtain an upper bound for the degree of g over $\mathbb{C}(f_1,\dots,f_n)$ independent of the individual g we have to alter our procedure slightly. We take into account only the functions $f_1, \dots, f_n$ when satisfying conditions (a)–(d). Then the bound given by (8) depends only on $f_1,\dots,f_n$. Let g be any function in $K(\Gamma)$. We already know that g satisfies an algebraic equation over $\mathbb{C}(f_1,\dots,f_n)$ and we want to derive local representations for g from there. Indeed, if m is the degree of the normed algebraic equation for g over $\mathbb{C}(f_1,\dots,f_n)$, and if $p(f_1,\dots,f_n)$ is a common denominator of the coefficients, multiply this equation by the mth power of p; then $g \cdot p(f_1,\dots,f_n)$ is seen to be integral over the polynomial ring $\mathbb{C}[f_1,\dots,f_n]$. Using the fact that the ring of holomorphic functions is integrally closed in the field of meromorphic functions, we can derive a local representation for $g \cdot p(f_1,\dots,f_n)$ as the quotient of holomorphic functions on $\bar{U}_\nu$, the denominators of which are products of the denominators of $f_1, \dots, f_n$. Then conditions (a)–(d) are automatically satisfied by $g \cdot p(f_1,\dots,f_n)$ as well. So $g \cdot p(f_1,\dots,f_n)$ and all the more g have bounded degree over $\mathbb{C}(f_1,\dots,f_n)$.

Remark
We cannot prove coprimality of the local representations of $g \cdot p(f_1, \ldots, f_n)$ obtained in this way. Therefore it was helpful to avoid the demand for coprimality in (a)–(d).

We have already drawn the main conclusions from Theorem 1, provided it may be applied to Siegel's modular group: Siegel modular functions of degree n (in the sense of Definition 3) form an algebraic function field of transcendence degree $n(n+1)/2$, and each modular function can be represented globally as the quotient of modular forms of equal weight. But we are still left with the proof that Siegel's modular group Γ_n is pseudoconcave for $n > 1$. We turn to this problem now. First we state a criterion for the pseudoconcavity of a boundary point.

Proposition 1
Let A be an open subset of $\mathbb{C}^n$, $n \geq 2$ and z^* a boundary point of A. Assume the existence of a complex plane

$$E = \{z = z^* + c_1 t_1 + c_2 t_2 \,|\, t_1, t_2 \in \mathbb{C}\}$$

($c_1, c_2 \in \mathbb{C}^n$ and linearly independent over $\mathbb{C}$) and an open neighborhood $V(0)$ of the origin in the t-space and a real-valued function $p(t_1, t_2, \bar{t}_1, \bar{t}_2)$ twice continuously differentiable on $V(0)$ such that

(i) *the Hermitian form $(\partial^2 p/\partial t_k \partial \bar{t}_l(0))$ is positive definite,*

(ii) $\{z^* + c_1 t_1 + c_2 t_2 | t_1, t_2 \in V(0), p(t_1, \ldots, \bar{t}_2) > p(0)\} \subset A.$

Then z^* is a pseudoconcave boundary point of A.

Remark
The proposition is still valid if the complex plane is replaced by an affine space of complex dimension $k \geq 2$ since then we may consider a two-dimensional affine subspace.

Proof
We set

$$p(t_1, \ldots, \bar{t}_2) = p(0) + \sum a_\nu t_\nu + \sum \bar{a}_\nu \bar{t}_\nu + \sum b_{\nu\mu} t_\nu t_\mu + \sum \bar{b}_{\nu\mu} \bar{t}_\nu \bar{t}_\mu + \sum c_{\nu\mu} t_\nu \bar{t}_\mu + q(t_1, \ldots, \bar{t}_2),$$

where ν, μ range over $\{1, 2\}$ and the coefficients are determined such that the derivatives of q up to order two vanish at the origin. Then in particular

$$c_{\nu\mu} = \frac{\partial^2 p}{\partial t_\nu \partial \bar{t}_\mu}(0),$$

and these numbers form the entries of a two-rowed positive definite matrix by condition (i). Consider the set

$$F = \left\{ z^* + c_1 t_1 + c_2 t_2 \,|\, \sum a_\nu t_\nu + \sum b_{\nu\mu} t_\nu t_\mu = 0, \quad t_1, t_2 \in V' \right\},$$

where $V'(0)$ is a sufficiently small open neighborhood of the origin contained in $V(0)$. Then, by conditions (i), (ii) and Taylor's theorem,

$$F - \{z^*\} \subset A, \tag{9}$$

whereas z^* belongs to F but not to A. If a_1 or a_2 is different from zero F is a one-dimensional complex manifold, if $a_1 = a_2 = 0$ and at least one $b_{\nu\mu} \neq 0$ we obtain two complex lines, and if all the a_ν and the $b_{\nu\mu}$ vanish we get a certain two-dimensional part of E. In any case the complex dimension is at least one. Take now an arbitrary sufficiently small open neighborhood U of z^*. Choose any open neighborhood U' of z^*, which lies relatively compact in U. Then (9) implies

$$\partial U' \cap F \subset U \cap A, \tag{10}$$

where $\partial U'$ denotes the boundary of U'. Apply the maximum principle to $U' \cap F$; then

$$|f(z^*)| \leq \sup_{\partial U' \cap F} |f|$$

holds for any holomorphic function f on U. We want to generalize this property by blowing up the underlying sets. For this purpose use the following notation. Let S be any subset of $\mathbb{C}^n$ and ε any positive real number. Then we denote by S_ε the set of all points in $\mathbb{C}^n$ whose distance to S is less that ε. Now since U' is relatively compact and $U \cap A$ is open, (10) implies

$$(\partial U' \cap F)_\varepsilon \subset U \cap A$$

for sufficiently small positive ε. Therefore, again by the maximum principle, we obtain

$$|f(z)| \leq \sup_{(\partial U' \cap F)_\varepsilon} |f| \leq \sup_{U \cap A} |f|$$

for all $z \in \{z^*\}_\varepsilon$ and each holomorphic function f on U. Hence

$$\{z^*\}_\varepsilon \subset (\widehat{U \cap A})_U$$

and z^* turns out to be a pseudoconcave boundary point of A.

We now consider the sets

$$L_n(t) = \left\{ z \in H_n \,|\, |x_{kl}| < t, \quad y \in Q_n'(t), \quad y_1 > \frac{1}{t} \right\}$$

introduced in Definition 3.2. Choose t so large that $L_n(t)$ contains Siegel's fundamental domain and write $L = L_n(t)$ for short. Then L contains at least one point of any Γ_n-orbit of H_n, and the set

$$\{m \in \Gamma_n \mid m\langle L\rangle \cap L \neq \varnothing\} \tag{11}$$

is finite by Theorem 3.1. We are interested in modular substitutions which act non-trivially at 'infinity' in the following sense.

Definition 4

An element m of Siegel's modular group is called a modular substitution at infinity, if

$$m\langle L - K\rangle \cap (L - K) \neq \varnothing \tag{12}$$

for all compact subsets K of H_n.

For instance, $m = \mathbf{1}$ is a modular substitution at infinity since L is not contained in any compact subset of H_n. Then we can say that obviously either both m and m^{-1} are modular substitutions at infinity or neither of them is.

Proposition 2

Let $m = \begin{pmatrix} a & b \\ c & d \end{pmatrix}$ be any modular substitution at infinity; then the last row and column of c vanish.

Proof

Since also m^{-1} is a modular substitution at infinity it is sufficient to consider the last column of c. Choose in particular $K = \varnothing$ in Definition 4 to deduce $m^{-1}\langle L\rangle \cap L \neq \varnothing$ and let z range over this set. For any

$$z \in m^{-1}\langle L\rangle \cap L$$

put $z^* = m\langle z\rangle$. From $az + b = z^*(cz + d)$ we infer for the real parts that

$$y^*cy = x^*d - ax - b + x^*cx.$$

Since $z, z^* \in L$, the real parts x, x^* are bounded; hence the right-hand side is seen to be bounded with respect to z. Use the Jacobian decompositions

$$y = s[v], \qquad y^* = s^*[v^*]$$

as in §2, where s, s^* are diagonal and v, v^* are triangular matrices. Then we deduce immediately the boundedness of

$$s^*v^*c\,{}^t\!vs$$

and from there the boundedness of $c\,{}^t\!vs$ since $z^* \in L$. The last column of $c\,{}^t\!vs$ consists of the elements

$$s_n c_{1n}, \ldots, s_n c_{nn}, \tag{13}$$

where s_n is the last diagonal element of s. Hence, these quantities are bounded with respect to z. However,

$$\sup_{z \in m^{-1}\langle L \rangle \cap L} s_n = \infty. \tag{14}$$

Indeed, a condition $s_n \leq q$ with constant q singles out a subset of L whose closure K is compact in H_n. If the supremum in (14) were finite and equal to q, we would obtain

$$m^{-1}\langle L \rangle \cap L \subset K,$$

thus contradicting (12). From the boundedness of the elements in (13) with respect to $z \in m^{-1}\langle L \rangle \cap L$ and (14) we infer that

$$c_{1n} = \cdots = c_{nn} = 0.$$

Corollary

For any modular substitution m at infinity $\det(cz + d)$ *is independent of the variables* $z_{\nu n}$ $(\nu = 1, \ldots, n)$ *in the last column of z.*

In fact, the last row and column of c vanish by the proposition. Hence the rank r of c is less than n and we can determine non-singular matrices v and $w = \begin{pmatrix} w_0 & 0 \\ 0 & 1 \end{pmatrix}$ such that

$$c^* = vc\,{}^t w = \begin{pmatrix} c_1 & 0 \\ 0 & 0 \end{pmatrix}, \qquad d^* = vdw^{-1} = \begin{pmatrix} d_1 & d_2 \\ d_3 & d_4 \end{pmatrix}.$$

Here c_1, d_1 are supposed to be r-rowed whereas w_0 consists of $n - 1$ rows and columns. Since c_1 is non-singular and $c^*\,{}^t d^*$ is symmetric we infer that $d_3 = 0$. Hence,

$$\det(v(cz + d)w^{-1}) = \det(c^* z[w^{-1}] + d^*)$$

does not depend on the last column of z.

Next we study the function

$$k(z) = -\log \det y$$

on H_n. We want to show that the $(n(n + 1)/2)$-rowed Hermitian matrix with entries

$$\frac{\partial^2 k}{\partial z_{\kappa\lambda} \partial \bar{z}_{\rho\sigma}} \qquad (\kappa \leq \lambda, \rho \leq \sigma) \tag{15}$$

is positive definite everywhere on H_n. For this purpose we have to consider the corresponding Hermitian form in $n(n + 1)/2$ variables, which we put together in a complex symmetric matrix a. First we compute

$$\sum_{\kappa\leq\lambda} a_{\kappa\lambda}\frac{\partial}{\partial z_{\kappa\lambda}}\log\det y = \sum_{\kappa\leq\lambda} a_{\kappa\lambda}\det y^{-1}\frac{\partial}{\partial z_{\kappa\lambda}}\det y$$

$$= \frac{1}{2\mathrm{i}}\sum_{\kappa,\lambda} a_{\kappa\lambda}\det y^{-1} y^*_{\kappa\lambda} = \frac{1}{2\mathrm{i}}\sigma(ay^{-1}),$$

where y^* is the adjoint of y and σ denotes the trace. Then we obtain for the Hermitian form in question

$$\sum_{\kappa\leq\lambda}\sum_{\rho\leq\sigma} a_{\kappa\lambda}\bar{a}_{\rho\sigma}\frac{\partial^2 k}{\partial z_{\kappa\lambda}\partial\bar{z}_{\rho\sigma}} = -\sum_{\rho\leq\sigma}\bar{a}_{\rho\sigma}\frac{\partial}{\partial\bar{z}_{\rho\sigma}}\left(\sum_{\kappa\leq\lambda} a_{\kappa\lambda}\frac{\partial}{\partial z_{\kappa\lambda}}\log\det y\right)$$

$$= -\frac{1}{2\mathrm{i}}\sigma\left(a\sum_{\rho\leq\sigma}\bar{a}_{\rho\sigma}\frac{\partial}{\partial\bar{z}_{\rho\sigma}}y^{-1}\right)$$

$$= \frac{1}{2\mathrm{i}}\sigma\left(ay^{-1}\sum_{\rho\leq\sigma}\bar{a}_{\rho\sigma}\left(\frac{\partial}{\partial\bar{z}_{\rho\sigma}}y\right)y^{-1}\right)$$

$$= \frac{1}{4}\sigma(ay^{-1}\bar{a}y^{-1}),$$

which is obviously positive for any $a \neq 0$. Of course by the derivative of a matrix we mean the matrix of the derivatives of its elements; furthermore we have used the elementary formula

$$\frac{\mathrm{d}}{\mathrm{d}t}y^{-1} = -y^{-1}\left(\frac{\mathrm{d}}{\mathrm{d}t}y\right)y^{-1}$$

for the derivative of the inverse during this computation.

We form the new function

$$p(z) = \min_{m\in\Gamma_n} k(m\langle z\rangle)$$

defined on H_n. Now we have

$$k(m\langle z\rangle) = k(z) + 2\log|\det(cz+d)|,$$

and $|\det(cz+d)|$, depending only on m modulo the subgroup $C_{n,0}$ of integral modular substitutions, tends to infinity uniformly on compact subsets of H_n if m runs over $C_{n,0}\backslash\Gamma_n$ (cf. §5). Hence the minimum in the definition of $p(z)$ may be restricted locally to finitely many values of m. So p turns out to be continuous. In general the minimum of $k(m\langle z\rangle)$ is attained for an m such that $m\langle z\rangle$ belongs to Siegel's fundamental domain F_n, since any point in F_n has maximal height in its orbit. Since F_n is contained in L we may restrict the formation of the minimum to the finitely many values of m appearing in (11) if z belongs to L. We want to show that we may even restrict ourselves to those m in (11) which are modular substitutions at infinity if we cut off a certain part of L. In fact, assume m to be any element of the set (11) which is not a modular substitution at infinity. Then we have

$$m\langle L\rangle \cap L \neq \emptyset \qquad \text{but} \qquad m\langle L - K\rangle \cap (L - K) = \emptyset$$

for a suitable compact set $K = K(m)$. Choose r so large that these finitely many sets $K(m)$ are contained in

$$\{z \in H_n \mid k(z) \geq -r\}.$$

Then

$$T = \{z \in L \mid k(z) < -r\}$$

satisfies $T \subset L - K(m)$ and $m\langle T\rangle \cap T = \emptyset$. Hence the finitely many $m = m_1, \ldots, m_s$ in Γ_n satisfying $m\langle T\rangle \cap T \neq \emptyset$ are modular substitutions at infinity. Moreover, for $z \in T$ the minimum of $k(m\langle z\rangle)$ is attained for an m with $m\langle z\rangle \in L$ and we have

$$k(m\langle z\rangle) \leq k(z) < -r.$$

Hence $m\langle z\rangle \in T$ and m belongs to the finite set $\{m_1, \ldots, m_s\}$. So we may write

$$p(z) = \min_{\nu=1,\ldots,s} k(m_\nu\langle z\rangle)$$

for $z \in T$, and all the m_ν are modular substitutions at infinity. Note that

$$k(m_\nu\langle z\rangle) = k(z) + 2\log|\det(c_\nu z + d_\nu)|$$

and that the terms $\det(c_\nu z + d_\nu)$ are independent of the variables in the last column of z by the corollary of Proposition 2. Hence for $z \in T$ the function $p(z)$ is differentiable with respect to $z_{1n}, \ldots, z_{nn}$, and the derivatives of order two with respect to these variables are

$$\frac{\partial^2 p}{\partial z_{\kappa n}\partial \bar{z}_{\lambda n}} = \frac{\partial^2 k}{\partial z_{\kappa n}\partial \bar{z}_{\lambda n}} \qquad (\kappa \leq \lambda).$$

They form an $n \times n$ Hermitian matrix, which is a submatrix of the positive definite Hermitian matrix composed of all the elements (15) and which is therefore again positive definite. So $p(z)$ considered as a function of $z_{1n}, \ldots, z_{nn}$ is a candidate in the criterion for pseudoconcavity stated in Proposition 1.

In order to prove the pseudoconcavity of Siegel's modular group Γ_n for $n > 1$, recall Definition 2 and choose for A the subset

$$A = \{z \in L \mid p(z) > -r - 1\}$$

of H_n. From the beginning, take r so large that A is non-empty. $p(z)$ is continuous, hence A is open in H_n. The condition $p(z) > -r-1$ implies $k(z) > -r-1$; therefore $\det y$ is bounded from above and A lies relatively compact in H_n. We must show that any point z_0 of the closure $\bar{A}$ can be mapped by an element of Γ_n onto a point of A or a pseudoconcave boundary point of A. Determine $m \in \Gamma_n$ such that

$$p(m\langle z_0 \rangle) = p(z_0) = k(m\langle z_0 \rangle)$$

and $m\langle z_0 \rangle \in L$. Since p is continuous we have $p(m\langle z_0 \rangle) = p(z_0) \geq -r-1$. Now if $m\langle z_0 \rangle \in A$ we are through; otherwise

$$m\langle z_0 \rangle \in \partial A, \qquad k(m\langle z_0 \rangle) = -r-1,$$

and the point $z^* = m\langle z_0 \rangle$ belongs to $T \cap \partial A$. We now apply Proposition 1 to show that z^* is a pseudoconcave boundary point of A. As the n-dimensional affine subspace we take

$$E = \left\{ z^* + \sum_{\nu=1}^{n} t_\nu \varepsilon_\nu \middle| t_\nu \in \mathbb{C} \right\},$$

where ε_ν denotes the $n \times n$ matrix with 1 at the entries (ν, n) and (n, ν), and 0 elsewhere. Furthermore consider the real-valued function

$$p^*(t_1, \ldots, t_n; \bar{t}_1, \ldots, \bar{t}_n) = p\left(z^* + \sum_{\nu=1}^{n} t_\nu \varepsilon_\nu \right)$$

in a sufficiently small neighborhood $V(0)$ of the origin in the t-space. Then from our former results on p we immediately deduce that p^* is twice continuously differentiable on $V(0)$ and that

$$\left(\frac{\partial^2 p^*}{\partial t_\kappa \partial \bar{t}_\lambda}(0) \right) = \left(\frac{\partial^2 p}{\partial z_{\kappa n} \partial \bar{z}_{\lambda n}}(z^*) \right)$$

is positive definite. Finally we verify that

$$\left\{ z = z^* + \sum_{\nu=1}^{n} t_\nu \varepsilon_\nu \middle| t \in V(0), p^*(t_1, \ldots, \bar{t}_n) > p^*(0) \right\} \subset A.$$

Indeed any element z in the set on the left belongs to L if $V(0)$ is sufficiently small, and the condition $p^*(t_1, \ldots, \bar{t}_n) > p^*(0)$ is equivalent to $p(z) > p(z^*) = -r-1$, so $z \in A$. Thus conditions (i) and (ii) of Proposition 1 are satisfied and z^* is proved to be a pseudoconcave boundary point of A.

We summarize our results and state

Theorem 2

Siegel's modular group Γ_n of arbitrary degree $n > 1$ is a pseudoconcave group of biholomorphic automorphisms of H_n.

VI

Dirichlet series

12 Dirichlet series associated with modular forms and the Mellin-transform

There are many connections between different kinds of Dirichlet series on the one hand and modular forms on the other. The extended work on this important subject is based on the ingenious ideas of the late Erich Hecke. We single out only one type of Dirichlet series and investigate their analytic continuations to meromorphic functions in the whole complex plane and their functional equations. The motivation for our selection comes from the method of proof, which will appear to be an application of our former theory of Eisenstein series (cf. §5). So this final chapter fits into this book very well, and the reader should excuse the narrowness of our point of view.

Let $f \in M_n^k$ be a modular form of degree n and even weight k, and let

$$f(z) = \sum_{t \geq 0} a(t) e^{2\pi i \sigma(tz)}$$

be its Fourier expansion. We consider the action of the unimodular group U_n on the space P_n of positive definite quadratic forms as in §2 and denote by

$$\varepsilon(t) = \#\{u \in U_n \,|\, t[u] = t\}$$

the number of units of $t \in P_n$. Then we associate with f the Dirichlet series

$$D_n(f, s) = \sum_{\{t\} > 0} \frac{a(t)}{\varepsilon(t)} \det t^{-s}, \tag{1}$$

where t runs over a complete set of half-integral positive matrices modulo the action of U_n. To investigate its convergence we need an estimate for the Fourier coefficients of f.

Lemma 1

The Fourier coefficients of any modular form $f(z)$ of weight k satisfy

$$|a(t)| \leq c \det t^k \qquad (t > 0),$$

where the constant c is independent of t.

Proof

We may assume $k > 0$. Let z be any point in H_n. Determine the modular substitutions

$$m = \begin{pmatrix} a & b \\ c & d \end{pmatrix}, \qquad m_0 = \begin{pmatrix} s & -1 \\ 1 & 0 \end{pmatrix}$$

such that $z_1 = m\langle z \rangle$ belongs to Siegel's fundamental domain, $\det(cs + d) \neq 0$ and $s - x$ is bounded. Such a choice is possible, since $\det(cs + d)$ is a polynomial in the entries of s, not identically zero and of degree at most two in each variable. Set $z_0 = m_0^{-1}\langle z \rangle$. Then we obtain by the cocycle relation

$$f(z) = j(mm_0, z_0)^{-k} j(m_0^{-1}, z)^{-k} f(z_1).$$

The first factor satisfies

$$|j(mm_0, z_0)|^{-k} = |\det((cs + d)z_0 - c)|^{-k} \leq \det y_0^{-k}$$

since $cs + d$ is non-singular and integral. Because of

$$\det y_0 = \det y \, |j(m_0^{-1}, z)|^{-2}$$

we infer that

$$|f(z)| \leq \det y^{-k} |j(m_0^{-1}, z)|^k |f(z_1)|.$$

Now

$$j(m_0^{-1}, z) = \det(-iy + s - x)$$

and $s - x$ is bounded. Hence this factor can be estimated by $(1 + \sigma(y))^n$ up to a constant factor. Furthermore $f(z_1)$ is bounded by Koecher's theorem. So we obtain

$$|f(z)| \leq c_1 (1 + \sigma(y))^{nk} \det y^{-k}, \tag{2}$$

where c_1 is independent of z. We apply this estimate to the formula

$$a(t) = \int_{x \bmod 1} f(x + it^{-1}) e^{-2\pi i \sigma((x + it^{-1})t)} \, dx \tag{3}$$

for the Fourier coefficients. Here we may assume t to be reduced because of (4.2). Furthermore apply Lemma 2.2 and replace t by its diagonalization to see that $\sigma(t^{-1})$ is bounded. Then the assertion follows from (2), (3) and these remarks.

The estimate of the Fourier coefficients in the lemma and Minkowski's reduction conditions imply immediately that the Dirichlet series (1) converges if the real part of s is sufficiently large.

H. Maass proved in [51] that $D_n(f, s)$ has an analytic continuation to a meromorphic function on $\mathbb{C}$ and that it satisfies a functional equation. For his proof he used the theory of invariant differential operators on weakly symmetric Riemannian spaces, which is technically rather involved. Later T. Arakawa gave another simpler proof in [3], and he even derived a residue-formula for the analytic continuation. We shall present Arakawa's proof in the following section.

The function given by (1) can be derived from the modular form $f(z)$ by another procedure, which is called the Mellin-transform of f. In order to prove the existence of the underlying improper integrals the following lemma will be useful.

Lemma 2

Let $\Phi(t)$ be any (real-valued) continuous function on the positive real line and $0 < y_1 < y_2$. Then

$$\int_{R_n, y_1 < \det y < y_2} \Phi(\det y) \det y^{-(n+1)/2}\, dy = \frac{n+1}{2} v_n \int_{y_1}^{y_2} \Phi(t) t^{-1}\, dt, \tag{4}$$

where R_n denotes Minkowski's reduced domain and

$$v_n = \int_{R_n, \det y \le 1} dy$$

is the Euclidean volume of that part of R_n indicated above.

Proof

The existence of the volume v_n is a simple consequence of Minkowski's reduction theory and a special case of Lemma 5.4. From there we immediately deduce the existence of the improper integral on the left-hand side of (4). Because of homogeneity we have

$$\int_{R_n, \det y \le t} dy = v_n\, t^{(n+1)/2}. \tag{5}$$

In order to prove (4) we look for a primitive of $\Phi(t)t^{-1}$. Define

$$f(t) = \int_{R_n, t_0 < \det y < t} \Phi(\det y) \det y^{-(n+1)/2}\, dy \qquad (t > t_0),$$

where t_0 is fixed, $0 < t_0 < y_1$. Then by a well-known mean-value theorem and (5) we deduce

$$\begin{aligned} \frac{f(t+h) - f(t)}{h} &= \frac{1}{h} \int_{R_n, t < \det y < t+h} \Phi(\det y) \det y^{-(n+1)/2}\, dy \\ &= \frac{1}{h} \Phi(t + \Theta h)(t + \Theta h)^{-(n+1)/2} \int_{R_n, t < \det y < t+h} dy \\ &= \Phi(t + \Theta h)(t + \Theta h)^{-(n+1)/2} v_n \frac{(t+h)^{(n+1)/2} - t^{(n+1)/2}}{h}, \end{aligned}$$

where $0 \leq \Theta \leq 1$ and, for simplicity, h was assumed to be positive. So we obtain

$$f'(t) = \frac{n+1}{2} \mathrm{v}_n \Phi(t) t^{-1},$$

and we have found a primitive for the integrand on the right-hand side of (4). The assertion follows from the fundamental theorem of integral calculus.

If f is a modular form of degree n and weight k, we consider that part

$$f^*(z) = \sum_{t>0} a(t) \mathrm{e}^{2\pi i \sigma(tz)} \tag{6}$$

of its Fourier expansion in which the summation runs only over all positive definite half-integral t. The Mellin-transform of f is defined by the improper integral

$$\xi_n(f, s) = \int_{R_n} \det y^s f^*(\mathrm{i}y)\, \mathrm{dv}. \tag{7}$$

Here

$$\mathrm{dv} = \det y^{-(n+1)/2}\, \mathrm{d}y$$

denotes the volume element on P_n which is invariant with respect to the action of $GL(n, \mathbb{R})$ (cf. §2).

Proposition 1

The Mellin-transform exists if $\operatorname{Re} s > k + (n+1)/2$.

Proof

Using Lemma 1 we obtain the following trivial estimate for the integrand in (7)

$$\begin{aligned} |\det y^s f^*(\mathrm{i}y)| &\leq c \det y^{\operatorname{Re} s} \sum_{t>0} \det t^k \mathrm{e}^{-2\pi\sigma(ty)} \\ &\leq c_1 \det y^{\operatorname{Re} s - k} \sum_{t>0} \mathrm{e}^{-\pi\sigma(ty)}, \end{aligned}$$

where $c_1, c_2, \ldots$ do not depend on s or y. If y is reduced in the sense of Minkowski, we have

$$\sum_{t>0} \mathrm{e}^{-\pi\sigma(ty)} \leq c_2 \det y^{-(n+1)/2} \mathrm{e}^{-c_3 \det y^{1/n}}.$$

This elementary inequality can be checked easily using Lemma 2.2 and replacing y by its diagonalization. So the Mellin-transform satisfies

$$|\xi_n(f,s)| \leq c_4 \int_{R_n} \det y^{\operatorname{Re} s - k - (n+1)/2} e^{-c_3 \det y^{1/n}} \mathrm{d}v.$$

We apply Lemma 2 to this integral, setting

$$\Phi(t) = t^{\operatorname{Re} s - k - (n+1)/2} e^{-c_3 t^{1/n}},$$

and obtain

$$|\xi_n(f,s)| \leq c_5 \int_0^\infty t^{\operatorname{Re} s - k - (n+1)/2 - 1} e^{-c_3 t^{1/n}} \mathrm{d}t$$

$$= c_5 n \int_0^\infty t^{n(\operatorname{Re} s - k - (n+1)/2) - 1} e^{-c_3 t} \mathrm{d}t.$$

Hence the question of convergence is finally reduced to a Eulerian integral of the second kind. It is finite for

$$\operatorname{Re} s > k + \frac{n+1}{2}.$$

Remarks

(i) By the proof, the convergence is uniform with respect to s on compact sets. Hence the Mellin-transform represents a holomorphic function on $\operatorname{Re} s > k + (n+1)/2$.

(ii) The partial integral

$$\int_{R_n,\, \det y \geq 1} \det y^s f^*(\mathrm{i}y)\, \mathrm{d}v$$

represents an entire function of s, since in that case we are reduced to a Eulerian integral of the second kind extended from one to infinity, which is holomorphic everywhere on $\mathbb{C}$.

(iii) We have used only the existence of a Fourier expansion of type (6) together with the estimate for the Fourier coefficients stated in Lemma 1. Hence f^* does not necessarily have to come from a modular form.

The connection between the Dirichlet series (1) and the Mellin-transform (7) is stated in

Proposition 2

Let $f(z)$ be any modular form of even weight k. Then the equation

$$\xi_n(f,s) = 2(2\pi)^{-ns} \prod_{r=1}^{n} \pi^{(r-1)/2} \Gamma\left(s - \frac{r-1}{2}\right) D_n(f,s)$$

holds for all complex s whose real part is sufficiently large.

Proof

We start with the definition (7) of the Mellin-transform, insert the Fourier expansion (6), and integrate term by term:

$$\begin{aligned}\xi_n(f,s) &= \int_{R_n} \det y^s \sum_{t>0} a(t) e^{-2\pi\sigma(ty)}\, dv \\ &= \sum_{t>0} a(t) \int_{R_n} e^{-2\pi\sigma(ty)} \det y^s\, dv \\ &= \sum_{\{t\}>0} \frac{a(t)}{\varepsilon(t)} \sum_{u\in U_n} \int_{R_n} e^{-2\pi\sigma(t[u]y)} \det y^s\, dv \\ &= 2 \sum_{\{t\}>0} \frac{a(t)}{\varepsilon(t)} \int_{P_n} e^{-2\pi\sigma(ty)} \det y^s\, dv.\end{aligned}$$

The integral on the right was evaluated in Lemma 6.2. From there we get

$$\xi_n(f,s) = 2(2\pi)^{-ns} \prod_{r=1}^{n} \pi^{(r-1)/2}\Gamma\left(s - \frac{r-1}{2}\right) \sum_{\{t\}>0} \frac{a(t)}{\varepsilon(t)} \det t^{-s}.$$

It is easily verified that summation and integration may be interchanged.

13 Analytic continuation and the functional equation

In this section we present T. Arakawa's proof for the analytic continuation of the Dirichlet series $D_n(f,s)$ associated with a modular form f of weight k and for its functional equation. If $k > 2n$ and even the linear spaces M_n^k are spanned by the corresponding Eisenstein series according to Proposition 5.6. Hence we may first restrict ourselves to those Dirichlet series which are attached to Eisenstein series. The generalization to arbitrary modular forms will then be a simple consequence of our former results in §5. Since Eisenstein series are of a rather explicit nature, we may expect some advantage from this procedure.

By our results in the previous section we can consider the Mellin-transform of the Eisenstein series $E_{n,r}^k(*;f)$ as well. But then we have to single out that part of its Fourier expansion in which the summation is taken over all positive definite half-integral t. We call the corresponding function

$$E_{n,r}^{k*}$$

in accordance with the notation in (12.6). Our first aim is the characterization of $E_{n,r}^{k*}$ as a subseries of the original Eisenstein series introduced in Definition 5.2.

For this purpose we have to introduce certain subgroups and subsets of Siegel's modular group Γ_n and the unimodular group U_n. In the following we denote by $\mathbf{1}_\nu$ the identity matrix of size ν.

Definition

Let $0 \le r \le \mu \le n$ be integers. Then

$$U_{\mu,n-\mu} = \left\{ u \in U_n \,\middle|\, u = \begin{pmatrix} u_1 & u_2 \\ 0 & u_4 \end{pmatrix}, \text{ where } u_1 \text{ is of size } \mu \right\},$$

$$\Gamma_n^{r,\mu} = \{ m \in \Gamma_n \mid \operatorname{rank}(0 \;\; \mathbf{1}_{n-r})c = \mu - r \},$$

$$\Gamma_n^r = \Gamma_n^{r,n}.$$

Note that the $U_{\mu,n-\mu}$ are subgroups of U_n, whereas the $\Gamma_n^{r,\mu}$ are subsets of Γ_n consisting of full left cosets of Γ_n modulo $C_{n,r}$. With this notation we form the set

$$L_n^{r,\mu} = \left\{ \left(\begin{array}{cc|cc} a_1 & 0 & b_1 & 0 \\ 0 & \mathbf{1}_{n-\mu} & 0 & 0 \\ \hline c_1 & 0 & d_1 & 0 \\ 0 & 0 & 0 & \mathbf{1}_{n-\mu} \end{array} \right) \begin{pmatrix} {}^t u & 0 \\ 0 & u^{-1} \end{pmatrix} \middle| \, m_1 \in C_{\mu,r} \backslash \Gamma_\mu^r, \;\; u \in U_n / U_{\mu,n-\mu} \right\}.$$

Then the disjoint union

$$\bigcup_{\mu=r}^{n} L_n^{r,\mu} \tag{1}$$

turns out to be a complete set of representatives of the left cosets of Γ_n modulo $C_{n,r}$. The proof of this fact is elementary but tedious. We omit it here and refer to [3] for the details. So the set (1) can serve to sum the Eisenstein series $E_{n,r}^k$.

Next we decompose the Fourier expansion of any modular form g of even weight into

$$g(z) = \sum_{\mu=0}^{n} g_\mu(z),$$

where

$$g_\mu(z) = \sum_{\operatorname{rank} t = \mu} a(t) e^{2\pi i \sigma(tz)}.$$

So in particular $g_n = g^*$. Now

$$t = \begin{pmatrix} t_1 & 0 \\ 0 & 0 \end{pmatrix} [{}^t u]$$

runs over all half-integral positive semi-definite matrices of size n and rank μ if t_1 ranges over all half-integral positive definite matrices of size μ and

u ranges over $U_n/U_{\mu,n-\mu}$. So we get

$$g_\mu(z) = \sum_{u \in U_n/U_{\mu,n-\mu}} \sum_{t_1>0} a\begin{pmatrix} t_1 & 0 \\ 0 & 0 \end{pmatrix} \exp 2\pi i\sigma\left(t_1 z\left[u\begin{pmatrix} \mathbf{1}_\mu \\ 0 \end{pmatrix}\right]\right).$$

If we compare this formula with how Siegel's Φ-operator affects the Fourier expansion (cf. (5.1)), we obtain

$$g_\mu(z) = \sum_{u \in U_n/U_{\mu,n-\mu}} (g|\Phi^{n-\mu})^*\left(z\left[u\begin{pmatrix} \mathbf{1}_\mu \\ 0 \end{pmatrix}\right]\right).$$

We apply this formula to the Eisenstein series $E^k_{n,r}(*,f)$, where $k > n+r+1$, $k \equiv 0 \bmod 2$ and f is a cusp form in S^k_r. In that case

$$E^k_{n,r}(*;f)|\Phi^{n-\mu} = \begin{cases} E^k_{\mu,r}(*;f) & \text{for } r \le \mu \le n \\ 0 & \text{for } \mu < r \end{cases}$$

by (5.17). Hence after taking the sum over μ we obtain

$$E^k_{n,r}(z;f) = E^{k*}_{n,r}(z;f) + \sum_{\mu=r}^{n-1} \sum_{u \in U_n/U_{\mu,n-\mu}} E^{k*}_{\mu,r}\left(z\left[u\begin{pmatrix} \mathbf{1}_\mu \\ 0 \end{pmatrix}\right];f\right). \qquad (2)$$

From this formula we deduce the following characterization of $E^{k*}_{n,r}$ as a subseries of the original Eisenstein series:

$$E^{k*}_{n,r}(z;f) = \sum_{m \in C_{n,r}\backslash\Gamma^r_n} f\left(m\langle z\rangle\begin{bmatrix} \mathbf{1}_r \\ 0 \end{bmatrix}\right) j(m,z)^{-k} \qquad (3)$$

($k > n+r+1$, k even). Indeed, we may argue by induction with respect to n if r is fixed. Assume (3) to be valid for all μ with $r \le \mu < n$ instead of n; then by the definition of $L^{r,\mu}_n$ we can rewrite (2) as

$$E^k_{n,r}(z;f) = E^{k*}_{n,r}(z;f) + \sum_{\mu=r}^{n-1} \sum_{m \in L^{r,\mu}_n} f\left(m\langle z\rangle\begin{bmatrix} \mathbf{1}_r \\ 0 \end{bmatrix}\right) j(m,z)^{-k}.$$

Since m ranges over the set (1) in the original Eisenstein series, $E^{k*}_{n,r}(z;f)$ consists of those terms of it for which m belongs to $L^{r,n}_n = C_{n,r}\backslash\Gamma^r_n$. Finally we point out that $E^{k*}_{n,r}(*;f)$ is not a modular form but behaves like a modular form only with respect to all integral modular substitutions.

Besides the subseries (3) of $E^k_{n,r}$ we consider another one defined by

$$P^k_{n,r}(z;f) = \sum_{m \in C_{n,r}\backslash\tilde{\Gamma}^r_n} f\left(m\langle z\rangle\begin{bmatrix} \mathbf{1}_r \\ 0 \end{bmatrix}\right) j(m,z)^{-k} \qquad (4)$$

($k > n+r+1$, k even), where now

$$\tilde{\Gamma}^r_n = \{m \in \Gamma_n | \operatorname{rank}(0\ \mathbf{1}_{n-r})c = \operatorname{rank}(0\ \mathbf{1}_{n-r})d = n-r\}.$$

This set $\tilde{\Gamma}^r_n$ is invariant if multiplied by

$$\iota = \begin{pmatrix} 0 & 1 \\ -1 & 0 \end{pmatrix} \quad \text{or} \quad \begin{pmatrix} {}^t u & 0 \\ 0 & u^{-1} \end{pmatrix} \qquad (u \in U_n) \tag{5}$$

from the right and consists again of full left cosets of Γ_n modulo $C_{n,r}$. Hence $P^k_{n,r}(z;f)$ satisfies

$$P^k_{n,r}(-z^{-1};f) = \det z^k P^k_{n,r}(z;f), \qquad P^k_{n,r}(z[u];f) = P^k_{n,r}(z;f)$$

and therefore behaves like a modular form with respect to the subgroup of Γ_n generated by the elements (5). In the following we shall use this behavior of $E^{k*}_{n,r}$ and $P^k_{n,r}$ with respect to different subgroups of the modular group rather frequently.

The fact that these two functions are nevertheless closely related is non-trivial and the content of

Proposition 1

Let $0 \le r \le n$, $k > n + r + 1$ and even, and f a cusp form in S^k_r. Then

$$E^{k*}_{n,r}(z;f) = P^k_{n,r}(z;f) + \det z^{-k} \sum_{\mu=r}^{n-1} \sum_{u \in U_n/U_{n-\mu,\mu}} P^k_{\mu,r}\left(-(z[u])^{-1}\begin{bmatrix} 0 \\ 1_\mu \end{bmatrix}; f\right)$$

and the series on the right converges absolutely and uniformly on compact subsets of H_n.

Proof

Similarly to $L^{r,\mu}_n$ we define for $0 \le r \le \mu \le n$

$$K^{r,\mu}_n = \left\{ \left(\begin{array}{cc|cc} a_1 & 0 & b_1 & 0 \\ 0 & 1_{n-\mu} & 0 & 0 \\ \hline c_1 & 0 & d_1 & 0 \\ 0 & 0 & 0 & 1_{n-\mu} \end{array}\right) \begin{pmatrix} {}^t u & 0 \\ 0 & u^{-1} \end{pmatrix} \middle| m_1 \in C_{\mu,r}\backslash\Gamma^r_\mu, \; u \in U_n/U_{\mu,n-\mu} \right\}.$$

Then the disjoint union

$$\bigcup_{\mu=r}^{n} K^{r,\mu}_n$$

turns out to be a complete set of representatives of the left cosets of $\Gamma^r_n \iota$ modulo $C_{n,r}$. Hence we obtain from (3)

$E^{k*}_{n,r}(z;f)$

$$= \sum_{m \in C_{n,r}\backslash\Gamma^r_n \iota} f\left(m\,\iota\langle z\rangle \begin{bmatrix} 1_r \\ 0 \end{bmatrix}\right) j(m\,\iota, z)^{-k}$$

$$= P^k_{n,r}(z;f) + \det z^{-k} \sum_{\mu=r}^{n-1} \sum_{m \in K^{r,\mu}_n} f\left(m\langle -z^{-1}\rangle \begin{bmatrix} 1_r \\ 0 \end{bmatrix}\right) j(m, -z^{-1})^{-k}. \tag{6}$$

On the other hand we deduce from (4) and the transformation properties of $P^k_{n,r}$:

$$\sum_{u \in U_n/U_{n-\mu,\mu}} P^k_{\mu,r}\left(-(z[u])^{-1}\begin{bmatrix}0\\ \mathbf{1}_\mu\end{bmatrix};f\right)$$

$$= \sum_{u \in U_n/U_{\mu,n-\mu}} P^k_{\mu,r}\left(-(z^{-1}[u])\begin{bmatrix}\mathbf{1}_\mu\\ 0\end{bmatrix};f\right)$$

$$= \sum_{m \in K_n^{r,\mu}} f\left(m\langle -z^{-1}\rangle\begin{bmatrix}\mathbf{1}_r\\ 0\end{bmatrix}\right) j(m, -z^{-1})^{-k}. \tag{7}$$

The assertion follows from (6) and (7). The question of convergence is already settled because of our former results on Eisenstein series in §5.

After these preparations we study the Mellin-transform of the Eisenstein series with regard to its analytic continuation and functional equation. We introduce the numerical constants

$$q(n) = \begin{cases} \pi^{1/2-n(n+1)/4} \prod_{\nu=2}^{n} \zeta(\nu)\Gamma\left(\dfrac{\nu}{2}\right) & \text{for } n \geq 2 \\ 1 & \text{for } n = 1, \end{cases}$$

where ζ denotes Riemann's ζ-function. They are related by

$$\mathrm{v}_n = \frac{2}{n+1} q(n) \tag{8}$$

to the Euclidean volume of that part of R_n for which $\det y \leq 1$ (cf. Lemma 12.2). Moreover we set $\varepsilon_{nk} = (-1)^{nk/2}$.

Proposition 2

Let $1 \leq r \leq n$, $k > n + r + 1$ and even, and $f \in S_r^k$. Then

(i) *the Mellin-transform $\xi_n(E^k_{n,r}(*;f), s)$ has an analytic continuation to a meromorphic function on $\mathbb{C}$;*

(ii) $$\xi_n(E^k_{n,r}(*;f), s) = \int_{R_n, \det y \geq 1} (\det y^s + \varepsilon_{nk} \det y^{k-s}) E^{k*}_{n,r}(iy;f)\, d\mathrm{v}$$

$$+ \frac{1}{2} \sum_{\mu=r}^{n-1} q(n-\mu)\xi_\mu\left(E^k_{\mu,r}(*;f), \frac{n}{2}\right)\left(\frac{\varepsilon_{nk}}{s-k+\mu/2} - \frac{1}{s-\mu/2}\right);$$

(iii) *for* $\dfrac{n-1}{2} < \operatorname{Re} s < k - \dfrac{n-1}{2}$

$$\xi_n(E^k_{n,r}(*;f), s) = \int_{R_n} \det y^s P^k_{n,r}(iy;f)\, d\mathrm{v},$$

where the integral on the right is absolutely convergent.

Note first that the integral in (ii) represents an entire function of s because of Remark (ii) following Proposition 12.1. Then we mention that the terms

$$\xi_\mu\left(E^k_{\mu,r}(*;f),\frac{n}{2}\right)$$

in (ii) may be understood either as values of the analytic continuation of the Mellin-transform according to (i) or by the integral representation (iii).

Proof

We argue by induction on n, where r is assumed to be fixed. If $n = r$ the Eisenstein series is the cusp form f itself. Then using the behavior of f with respect to $z \mapsto -z^{-1}$ we obtain

$$\begin{aligned}\xi_n(f,s) &= \int_{R_n,\,\det y \geq 1} \det y^s f(iy)\,dv + \int_{R_n,\,\det y \leq 1} \det y^s f(iy)\,dv\\ &= \int_{R_n,\,\det y \geq 1} \det y^s f(iy)\,dv + \varepsilon_{nk}\int_{R_n,\,\det y \geq 1} \det y^{k-s} f(iy)\,dv.\end{aligned}$$

Hence the Mellin-transform converges everywhere representing an entire function of s and satisfying (ii) and (iii).
Assume now the proposition to be valid for all μ with $r \leq \mu < n$ instead of n. Under this hypothesis we have to prove statements (i)–(iii) for n. First we show

$$\begin{aligned}&\int_{R_n,\,\det y \geq 1} \det y^{-s} \sum_{u \in U_n/U_{n-\mu,\mu}} P^k_{\mu,r}\left(i(y[u])^{-1}\begin{bmatrix}0\\ \mathbf{1}_\mu\end{bmatrix};f\right)dv\\ &\quad = \frac{1}{2}q(n-\mu)\frac{1}{s-\mu/2}\xi_\mu\left(E^k_{\mu,r}(*;f),\frac{n}{2}\right)\end{aligned} \tag{9}$$

for $r \leq \mu < n$ and $\operatorname{Re} s > \mu/2$, the integral on the left being absolutely convergent. Indeed, with regard to the convergence of the integral consider

$$\begin{aligned}&\int_{R_n,\,\det y \geq 1} \det y^{-\operatorname{Re} s} \sum_{u \in U_n/U_{n-\mu,\mu}} \left|P^k_{\mu,r}\left(i(y[u])^{-1}\begin{bmatrix}0\\ \mathbf{1}_\mu\end{bmatrix};f\right)\right|dv\\ &\quad = \int_{Q,\,\det y \geq 1} \det y^{-\operatorname{Re} s}\left|P^k_{\mu,r}\left(iy^{-1}\begin{bmatrix}0\\ \mathbf{1}_\mu\end{bmatrix};f\right)\right|dv;\end{aligned} \tag{10}$$

here Q may be any fundamental domain of $U_{n-\mu,\mu}$ in P_n, since the integrand is invariant with respect to the action of $U_{n-\mu,\mu}$. Take for instance

$$Q = \left\{y = \begin{pmatrix}p & 0\\ 0 & t\end{pmatrix}\left[\begin{pmatrix}1 & x\\ 0 & 1\end{pmatrix}\right] \middle|\, p \in R_{n-\mu},\, t \in R_\mu,\, x \in I\right\},$$

where

$$I = \{x \mid 0 \leq x_{11} \leq \tfrac{1}{2};\ -\tfrac{1}{2} \leq x_{kl} \leq \tfrac{1}{2} \text{ for } (k,l) \neq (1,1)\}.$$

Then

$$dv_y = \det p^{\mu/2} \det t^{-(n-\mu)/2}\, dv_p\, dv_t\, dx,$$

and we obtain for (10)

$$\frac{1}{2}\int_{R_{n-\mu}\times R_\mu,\, \det p \det t \geq 1} \det p^{-\operatorname{Re} s+\mu/2} \det t^{-\operatorname{Re} s-(n-\mu)/2} |P^k_{\mu,r}(\mathrm{i}t^{-1}; f)|\, dv_p\, dv_t.$$

The integration with respect to p can be carried through. Using Lemma 12.2, the numerical value (8), and the assumption $\operatorname{Re} s > \mu/2$, we get

$$\int_{R_{n-\mu},\, \det p \geq \det t^{-1}} \det p^{-\operatorname{Re} s+\mu/2}\, dv_p = \frac{1}{\operatorname{Re} s - \mu/2} q(n-\mu) \det t^{\operatorname{Re} s-\mu/2}.$$

So the integral (10) is equal to

$$\frac{1}{2}\frac{1}{\operatorname{Re} s - \mu/2} q(n-\mu) \int_{R_\mu} \det t^{n/2} |P^k_{\mu,r}(\mathrm{i}t; f)|\, dv_t$$

and this integral is finite by the induction hypothesis. Therefore, by the bounded convergence theorem, the calculation above can be repeated for the integral (9) itself instead of (10) and leads to the desired result. Similarly we conclude from the induction hypothesis

$$\begin{aligned} &\int_{R_n,\, \det y\geq 1} \det y^{-s} \sum_{u\in U_n/U_{\mu,n-\mu}} P^k_{\mu,r}\left(\mathrm{i}(y[u])\begin{bmatrix}1_\mu\\0\end{bmatrix}; f\right) dv \\ &\quad = \frac{1}{2} q(n-\mu)\frac{1}{s+\mu/2}\xi_\mu\left(E^k_{\mu,r}(*; f), \frac{n}{2}\right) \end{aligned} \tag{11}$$

for $r \leq \mu < n$ and $\operatorname{Re} s > -\mu/2$, the integral again being absolutely convergent.

Now decompose the Mellin-transform of the Eisenstein series $E^k_{n,r}(*; f)$ in the following manner:

$$\begin{aligned} \xi_n(E^k_{n,r}(*; f), s) = &\int_{R_n,\, \det y \geq 1} \{\det y^s + \varepsilon_{nk} \det y^{k-s}\} E^{k*}_{n,r}(\mathrm{i}y; f)\, dv \\ &+ \int_{R_n,\, \det y\geq 1} \det y^{-s}\{E^{k*}_{n,r}(\mathrm{i}y^{-1}; f) - \varepsilon_{nk} \det y^k E^{k*}_{n,r}(\mathrm{i}y; f)\}\, dv \end{aligned} \tag{12}$$

for $\operatorname{Re} s > k + (n+1)/2$. Proposition 1 yields

$$\begin{aligned} &E^{k*}_{n,r}(\mathrm{i}y^{-1}; f) - \varepsilon_{nk} \det y^k E^{k*}_{n,r}(\mathrm{i}y; f) \\ &\quad = \varepsilon_{nk} \det y^k \sum_{\mu=r}^{n-1} \sum_{u\in U_n/U_{\mu,n-\mu}} P^k_{\mu,r}\left(\mathrm{i}(y[u])\begin{bmatrix}1_\mu\\0\end{bmatrix}; f\right) \\ &\quad\quad - \sum_{\mu=r}^{n-1} \sum_{u\in U_n/U_{n-\mu,\mu}} P^k_{\mu,r}\left(\mathrm{i}(y[u])^{-1}\begin{bmatrix}0\\1_\mu\end{bmatrix}; f\right). \end{aligned}$$

Hence (9) and (11) imply

$$\int_{R_n,\,\det y\geq 1} \det y^{-s}\{E_{n,r}^{k*}(\mathrm{i}y^{-1};f) - \varepsilon_{nk}\det y^k E_{n,r}^{k*}(\mathrm{i}y;f)\}\,\mathrm{d}v \tag{13}$$

$$= \frac{1}{2}\sum_{\mu=r}^{n-1} q(n-\mu)\xi_\mu\left(E_{\mu,r}^k(*;f),\frac{n}{2}\right)\left(\frac{\varepsilon_{nk}}{s-k+\mu/2} - \frac{1}{s-\mu/2}\right)$$

for $\operatorname{Re} s > k - r/2$. Combine (12) and (13) to infer statements (i) and (ii) for n.

To prove (iii) we conclude from Proposition 1 and (9) that

$$\int_{R_n,\,\det y\geq 1} \det y^s P_{n,r}^k(\mathrm{i}y;f)\,\mathrm{d}v = \int_{R_n,\,\det y\geq 1} \det y^s E_{n,r}^{k*}(\mathrm{i}y;f)\,\mathrm{d}v$$

$$-\frac{1}{2}\sum_{\mu=r}^{n-1} q(n-\mu)\xi_\mu\left(E_{\mu,r}^k(*;f),\frac{n}{2}\right)\frac{\varepsilon_{nk}}{k-s-\mu/2}$$

for $\operatorname{Re} s < k-(n-1)/2$. By the behavior of $P_{n,r}^k$ with respect to $z\mapsto -z^{-1}$ we have

$$\int_{R_n,\,\det y\leq 1} \det y^s P_{n,r}^k(\mathrm{i}y;f)\,\mathrm{d}v = \varepsilon_{nk}\int_{R_n,\,\det y\geq 1} \det y^{k-s} P_{n,r}^k(\mathrm{i}y;f)\,\mathrm{d}v$$

$$= \varepsilon_{n,k}\int_{R_n,\,\det y\geq 1} \det y^{k-s} E_{n,r}^{k*}(\mathrm{i}y;f)\,\mathrm{d}v$$

$$-\frac{1}{2}\sum_{\mu=r}^{n-1} q(n-\mu)\xi_\mu\left(E_{\mu,r}^k(*;f),\frac{n}{2}\right)\frac{1}{s-\mu/2}$$

for $\operatorname{Re} s > (n-1)/2$. Adding these two equations and using (ii) we obtain statement (iii) for n. The absolute convergence of the integral in (iii) is a consequence of the corresponding property of the integral in (9). The proof of Proposition 2 is now complete.

The last proposition can be extended to the case $r = 0$. The only difference consists in part (ii). In that case we have for any $a\in\mathbb{C}$

$$\xi_n(E_{n,0}^k(*;a),s) = \int_{R_n,\,\det y\geq 1} (\det y^s + \varepsilon_{nk}\det y^{k-s})E_{n,0}^{k*}(\mathrm{i}y;a)\,\mathrm{d}v$$

$$+ q(n)a\left(\frac{\varepsilon_{nk}}{s-k} - \frac{1}{s}\right) + \frac{1}{2}\sum_{\mu=1}^{n-1} q(n-\mu)\xi_\mu\left(E_{\mu,0}^k(*;a),\frac{n}{2}\right)$$

$$\times\left(\frac{\varepsilon_{nk}}{s-k+\mu/2} - \frac{1}{s-\mu/2}\right).$$

We showed in §5 how M_n^k is built up from Eisenstein series and how Siegel's Φ-operator acts; in particular we refer to Theorem 5.2. If we combine these results with Proposition 2 we immediately obtain the final

Theorem

Let $k > 2n$, $k \equiv 0 \bmod 2$ *and* f *be a modular form of weight* k *and degree* n. *Then* $\xi_n(f,s)$ *has an analytic continuation to a meromorphic function on* $\mathbb{C}$, *realized explicitly as the right-hand side of*

$$\xi_n(f,s) = \int_{R_n,\,\det y \geq 1} (\det y^s + \varepsilon_{nk} \det y^{k-s}) f^*(iy)\, dv$$

$$+ q(n)(f|\Phi^n)\left(\frac{\varepsilon_{nk}}{s-k} - \frac{1}{s}\right) + \frac{1}{2}\sum_{\mu=1}^{n-1} q(n-\mu)\xi_\mu\left(f|\Phi^{n-\mu}, \frac{n}{2}\right)$$

$$\times \left(\frac{\varepsilon_{nk}}{s-k+\mu/2} - \frac{1}{s-\mu/2}\right).$$

Each term on the right is invariant up to the factor ε_{nk} if s is replaced by $k-s$. Hence we obtain the functional equation

$$\xi_n(f, k-s) = (-1)^{nk/2}\xi_n(f,s).$$

Bibliography

[1] Andreotti, A. and Grauert, H., 'Algebraische Körper von automorphen Formen', *Nachr. Akadem. Wiss. Göttingen* 1961, 39–48

[2] Andrianov, A.N., *Quadratic forms and Hecke operators*, Grundl. math. Wiss. 286, Springer-Verlag (1987)

[3] Arakawa, T., 'Dirichlet series corresponding to Siegel's modular forms', *Math. Ann.* 238 (1978), 157–73

[4] Baily, W.L., Jr., 'Satake's compactification of V_n', *Amer. J. Math.* 80 (1958), 348–64

[5] Baily, W.L., Jr. and Borel, A., 'Compactification of arithmetic quotients of bounded symmetric domains', *Ann. of Math.* 84 (1966), 442–528

[6] Behr, H., 'Über die endliche Definierbarkeit von Gruppen', *J. reine angew. Math.* 211 (1962), 116–22

[7] Behr, H., 'Eine endliche Präsentation der symplektischen Gruppe $Sp_4(\mathbb{Z})$', *Math. Z.* 141 (1975), 47–56

[8] Böcherer, S., 'Über die Fourier–Jacobi–Entwicklung Siegelscher Eisensteinreihen', *Math. Z.* 183 (1983), 21–46

[9] Böcherer, S., 'Über die Fourier–Jacobi–Entwicklung Siegelscher Eisensteinreihen II', *Math. Z.* 189 (1985), 81–110

[10] Borel, A., 'Les fonctions automorphes de plusieurs variables complex', *Bull. soc. math. France* 80 (1952), 167–82.

[11] Borel, A. *Introduction aux groupes arithmétiques*, Hermann, Paris (1969)

[12] Braun, H., 'Konvergenz verallgemeinerter Eisensteinscher Reihen', *Math. Z.* 44 (1939), 387–97

[13] Cartan, E., 'Sur les domaines bornés homogènes de l'espace de *n* variables complexes', *Abh. Math. Sem. Hans. Univ.* 11 (1936), 116–62

[14] Christian, U., 'Zur Theorie der Modulfunktionen *n*-ten Grades', *Math. Ann.* 133 (1957), 281–97

[15] Christian, U., 'Über Hilbert–Siegelsche Modulformen and Poincarésche Reihen', *Math. Ann.* 148 (1962), 257–307

[16] Earle, C.J., 'Some remarks on Poincaré series', *Comp. Math.* 21 (1969), 167–76

[17] Eichler, M., 'The basis problem for modular forms and the traces of the Hecke operators', *Proc. International Summer School, Antwerp, Modular Functions of one Variable* I, 75–151, Lecture Notes in Mathematics 320, Springer-Verlag, Berlin (1973)

[18] Eichler, M., 'Über die Anzahl der linear unabhängigen Siegelschen Modulformen von gegebenem Gewicht', *Math. Ann.* 213 (1975), 281–91

[19] Faltings, G., 'Arithmetische Kompaktifizierung des Modulraums der abelschen Varietäten', *Proc. Arbeitstagung Bonn 1984*, 321–83, Lecture Notes in Mathematics 1111, Springer-Verlag (1985)

[20] Freitag, E., 'Zur Theorie der Modulformen zweiten Grades', *Nachr. Akadem. Wiss. Göttingen* 1965, 151–7

[21] Freitag, E., 'Holomorphe Differentialformen zu Kongruenzgruppen der Siegelschen Modulgruppe', *Invent. math.* 30 (1975), 181–96

[22] Freitag, E., 'Stabile Modulformen', *Math. Ann.* 230 (1977), 197–211

[23] Freitag, E., 'Eine Bemerkung zu Andrianovs expliziten Formeln für die Wirkung der Heckeoperatoren auf Thetareihen', *Christoffel-Festschrift*, 336–51, Birkhäuser-Verlag, Basel (1981)

[24] Freitag, E., *Siegelsche Modulfunktionen*, Grundl. math. Wiss. 254, Springer-Verlag (1983)

[25] Gerstenhaber, M., 'On the algebraic structure of discontinuous groups', *Proc. AMS* 4 (1953), 745–50

[26] Godement, R., Différents exposés, *Fonctions automorphes, vol.* 1, Sém. H. Cartan 1957/58, Paris

[27] Götzky, F., 'Über eine zahlentheoretische Anwendung von Modulfunktionen zweier Veränderlicher', *Math. Ann.* 100 (1928), 411–37

[28] Gottschling, E., 'Explizite Bestimmung der Randflächen des Fundamentalbereiches der Modulgruppe zweiten Grades', *Math. Ann.* 138 (1959), 103–24

[29] Gottschling, E., 'Über Poincarésche Reihen und einen Fundamentalbereich diskontinuierlicher Gruppen', *Math. Ann.* 148 (1962), 125–46

[30] Hashimoto, K.-I., 'The dimension of the space of cusp forms on Siegel upper half plane of degree two (I)', *J. Fac. Sc. Univ. Tokyo Sec. IA* 30 (1983), 403–88

[31] Hua, L.K., *Harmonic analysis of functions of several complex variables in the classical domains*, Transl. Math. Monographs, Vol. 6, AMS (1963)

[32] Hua, L.K. and Reiner, I., 'On the generators of the symplectic modular group', *Transact. AMS* 65 (1949), 415–26

[33] Humphreys, J.E., 'Variations on Milnor's computation of $K_2\mathbb{Z}$', *Algebraic K-theory II* (*Seattle* 1972), 304–7, Lecture Notes in Mathematics 342, Springer-Verlag, Berlin (1973)

[34] Igusa, J.-I., 'On Siegel modular forms of genus two', *Amer. J. Math.* 84 (1962), 175–200

[35] Igusa, J.-I., 'On Siegel modular forms of genus two (II)', *Amer. J. Math.* 86 (1964), 392–412

[36] Igusa, J.-I., 'Modular forms and projective invariants', *Amer. J. Math.* 89 (1967), 817–55

[37] Igusa, J.-I., *Theta functions*, Grundl. math. Wiss. 194, Springer-Verlag (1972)

[38] Kirchheimer, F., 'Über explizite Präsentationen Hilbertscher Modulgruppen zu totalreellen Körpern der Klassenzahl eins', *J. reine angew. Math.* 321 (1981), 120–37

[39] Kirchheimer, F. and Wolfart, J., 'Explizite Präsentation gewisser Hilbertscher Modulgruppen durch Erzeugende und Relationen', *J. reine angew. Math.* 315 (1980), 139–73

[40] Klingen, H., 'Über die Erzeugenden gewisser Modulgruppen', *Nachr. Akadem. Wiss. Göttingen* 1956, 173–85

[41] Klingen, H., 'Charakterisierung der Siegelschen Modulgruppe durch ein endliches System definierender Relationen', *Math. Ann.* 144 (1961), 64–72

[42] Klingen, H., 'Über Poincarésche Reihen zur Siegelschen Modulgruppe', *Math. Ann.* 168 (1967), 157–70

[43] Klingen, H., 'Zum Darstellungssatz für Siegelsche Modulformen', *Math. Z.* 102 (1967), 30–43

[44] Klingen, H., 'Berichtigung dazu', *Math. Z.* 105 (1968), 399–400

[45] Klingen, H., 'Zur Struktur der Siegelschen Modulgruppe', *Math. Z.* 136 (1974), 169–78

[46] Klingen, H., 'Metrisierungstheorie und Jacobiformen', *Abh. Math. Sem. Univ. Hamburg* 57 (1987), 165–78

[47] Koecher, M., 'Zur Theorie der Modulformen *n*-ten Grades I', *Math. Z.* 59 (1954), 399–416

[48] Langlands, R.P., *On the functional equations satisfied by Eisenstein series*, Lecture Notes in Mathematics 544, Springer-Verlag, Berlin (1976)

[49] Maass, H., 'Über die Darstellung der Modulformen *n*-ten Grades durch Poincarésche Reihen', *Math. Ann.* 123 (1951), 125–51

[50] Maass, H., 'Die Multiplikatorsysteme zur Siegelschen Modulgruppe', *Nachr. Akadem. Wiss. Göttingen* 1964, 125–35

[51] Maass, H., *Siegel's modular forms and Dirichlet series*, Lectures Notes in Mathematics 216, Springer-Verlag, Berlin (1971)

[52] Minkowski, H., 'Diskontinuitätsbereich für arithmetische Äquivalenz', *J. reine angew. Math.* 129 (1905), 220–74

[53] Mumford, D., 'On the Kodaira dimension of the Siegel modular variety (algebraic geometry – open problems)', 348–75, Lecture Notes in Mathematics 997, Springer-Verlag, Berlin (1982)

[54] Pjateckij-Šapiro, I.I., 'Singular modular functions', *Izvestija Akad. Nauk. SSSR, Ser. mat.* 20 (1956), 53–98 (in Russian)

[55] Pjateckij-Šapiro, I.I., 'On a problem of E. Cartan', *Rap. Ac. Sc. URSS* 124, no. 2 (1959), 272–3 (in Russian)

[56] Pjateckij-Šapiro, I.I., *Automorphic functions and the geometry of classical domains*, Gordon and Breach, New York (1969)

[57] Poincaré, H., 'Mémoire sur les fonctions Fuchsiennes', *Acta Math.* 1 (1883), 193–294

[58] Raghavan, S., 'Singular modular forms of degree *s*', C.P. Ramanujam – A Tribute, *Studies in Math.* 8 (1978), 263–72, Tata Institute of Fundamental Research, Bombay

[59] Resnikoff, H.L., 'Automorphic forms of singular weight are singular forms', *Math. Ann.* 215 (1975), 173–93

[60] Satake, I., 'On the compactification of the Siegel space', *J. Indian Math. Soc.* 20 (1956), 259–81

[61] Serre, J.-P., *A course in arithmetic*, Graduate Texts in Mathematics, vol. 7, Springer-Verlag, Berlin, 2nd ed. (1978)

[62] Siegel, C.L., 'Einführung in die Theorie der Modulfunktionen *n*-ten Grades', *Math. Ann.* 116 (1939), 617–57

[63] Siegel, C.L., 'Einheiten quadratischer Formen', *Abh. Math. Sem. Hans. Univ.* 13 (1940), 209–39

[64] Siegel, C.L., 'Symplectic geometry', *Amer. J. Math.* 65 (1943), 1–86

[65] Siegel, C.L., 'Zur Theorie der Modulfunktionen *n*-ten Grades', *Comm. Pure Appl. Math.* 8 (1955), 677–81

[66] Tai, Y.-S., 'On the Kodaira dimension of the moduli space of abelian varieties', *Invent. math.* 68 (1982), 425–39

[67] Tsuyumine, S., 'On Siegel modular forms of degree three', *Amer. J. Math.* 108 (1986), 755–862

[68] van der Waerden, B.L., 'Die Reduktionstheorie der positiven quadratischen Formen', *Acta Math.* 96 (1956), 265–309

[69] Weissauer, R., *Stabile Modulformen und Eisensteinreihen*, Lecture Notes in Mathematics 1219, Springer-Verlag, Berlin (1986)

[70] Weyl, H., 'Theory of reduction for arithmetical equivalence I', *Transact. AMS* 48 (1940), 126–64

[71] Weyl, H., 'Theory of reduction for arithmetical equivalence II', *Transact. AMS* 51 (1942), 203–31

[72] Witt, E., 'Eine Identität zwischen Modulformen zweiten Grades', *Abh. Math. Sem. Hans. Univ.* 14 (1941), 323–37

Index